Wild Solutions

To Doug and Keira

To Mom and Melissa:
far apart in age; close together in spirit

Contents

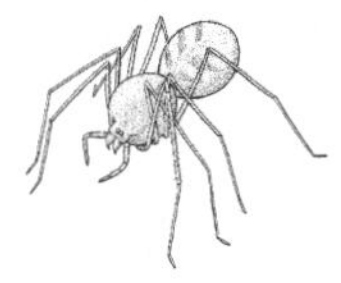

Preface

The discovery of new species—figuring out what they do in nature and how they contribute to our wealth and happiness—is one of the most exciting branches of science today. The study of biological diversity, or biodiversity for short, is revealing a previously unimagined variety of life forms on Earth. This research has become a powerful generator of ideas, opening fresh horizons of scientific knowledge with many new products and industries as spin-offs. It has also given us a glimpse into the workings of those natural mechanisms called ecosystems that keep us all alive.

As an example, how often have you heard a friend ask exasperatedly, "Why on earth do mosquitoes exist!?" as he swats the little animal with his hand. Yet we now know that adult mosquitoes are important components of the food chains that feed birds, and their larvae are a major ingredient of the diet of many fish. Most people appreciate birds and fish, one way or another, and therefore appreciate (with their heads, if not their hearts) the place of mosquitoes in nature. But there is more. Some orchids require mosquitoes for pollination; and research on one mosquito species in particular is revealing a potential breakthrough in the fight against malaria, one of the world's most disastrous diseases. We know all this because scientists are in the process of discovering just how many species of mosquito there are, what they do, and how they may be useful to us.

The study of biodiversity is providing answers to everyday questions about the millions of species that surround us—questions such as What do all these species do? What are they for? Do we need them all?

This book is for anyone interested in questions such as these, anyone interested in joining in the excitement of the discovery and exploration of life forms on this planet. The science of biodiversity is in its pioneering phase; we still know shockingly little about the species with which we share our tiny rock in space. Still, what we do know promises discoveries that will benefit all humanity. In many ways, this science resembles space exploration: both are in their infancy, and both are developing complex technologies to achieve their goals. However, in the science of biodiversity, microscopes replace telescopes, and the organisms we seek are on the ground beneath our feet rather than light years away.

The glimpses into the future offered in *Wild Solutions* show that our ignorance about the interactions of species in natural ecosystems demonstrates that we humans are not really in control. For example, many organisms regulate the fertility of the soil and the content of the atmosphere, but we have very little knowledge of which species are involved, how many there are, and what precisely they do. This book will also present many examples of the basic proposition that species that appear to be totally insignificant right now are likely in the future to become extremely valuable to medicine, to agriculture, and to a variety of other human needs. The species with which we share the planet are, if nothing else, a vast insurance policy against the problems we will probably face in the years to come.

Our ignorance also fosters a false view of the world. "Get real!" is a demand we often hear from people who wish to convince us that the harsh imperatives of modern life, especially economics and politics, must dominate decision-making in environmental and other debates. They claim that economics and politics are the real world. Yet the evidence that this hypothesis is false is an inescapable part of everyday living. Eating, drinking, and using the rest room remind us all too frequently that the real world is biological. Try ignoring your feelings the next time you are hungry or thirsty; your nervous system will soon forcibly remind you that you are an organism, and that your organs require food and drink on a regular basis. Or try ignoring your next fever. A host of microbes will soon make it clear that biology is what controls your life. While money enables you to buy food, drink, and medicine, it is biodiversity that provides these essentials in the first place.

We humans are a natural part of this biological world. Each one of us, every family and community, is a living, breathing inhabitant of an ecosystem. Every town, city, business, and industry is a wholly owned subsidiary of one or more ecosystems, each of which is shared with hundreds of thousands of other species. A tiny minority of these species are harmful, and unfortunately the media tend to focus on that handful. The vast majority of species are beneficial; if they were not out in the world, going about their daily business, life as we know it would cease.

The ecosystems in which we live and the many species on which we depend are the products of millions of years of evolution. The genetic information they contain is an irreplaceable source of information that has already provided humanity with all of its biological resources. In *Wild Solutions* we will show that ecosystems can provide us with a huge variety of new resources, products, and services—*if* we allow them to coexist with us, and *if* we manage them with foresight, imagination, enthusiasm, and respect.

We authors wish to thank our many colleagues who, over the years, have provided us with inspiration and knowledge. Their numbers are too great to be properly acknowledged here; however, we especially thank Dave Briscoe, Mark Dangerfield, Mike Gillings, Andy Holmes, and Noel Tait at Macquarie University, and Gretchen Daily, Anne Ehrlich, and the late Dick Holm at Stanford University. We are grateful to Robyn Delves, executive officer of the Commonwealth Key Centre for Biodiversity and Bioresources, who has assisted us in many ways. Also, Vivian Wheeler and Jean Thomson Black of Yale University Press have been sources of help and encouragement throughout the development and refinement of this book.

A.B.

P.E.

Wild Solutions

ONE
Introduction

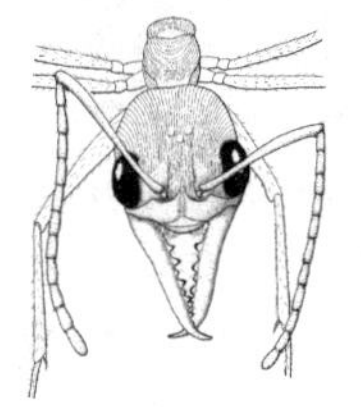

All we have yet discovered is but a trifle in comparison

with what lies hid in the great treasury of nature

—Antonie van Leeuwenhoek, 1708

In this book we explore what is known about the plants, animals, and microbes on Earth from the point of view of what they do for us: how they keep us alive, how they feed us, how they sustain our economy, how they are a source of immense wealth and well-being that we are still just beginning to understand. We will find that we depend on an extraordinary assortment of populations, species, and natural communities. This variety is known as biological diversity, or biodiversity for short.

The history of civilization is a history of human beings as they become increasingly knowledgeable about biological diversity. Over many centuries, men and women throughout the world have observed the variety of organisms around them and have identified thousands of useful species. Roman mythology depicted them as flowing from the horn of plenty, a ram's horn held by Ceres, the Mother Earth who was charged with the protection of agriculture and all it produced. Mythology the story may have been, but it was a clear recognition of the many species that made civilization possible and kept it supplied with everything people had needed for centuries.

Today, we are still dependent on much the same variety of animals and plants, both wild and domesticated. When asked to think of useful plant species, most of us can name a whole array: rice, corn, wheat, barley, oats, apples, bananas, plums, cabbage, broccoli, lettuces, carrots, onions, dates, rubber, quinine, peppers, cinnamon, oregano, roses, daffodils, orchids, oaks, pines, redwoods, coconuts, breadfruit, potatoes, watercress, lotus, seaweeds, grapes, peas, beans, coffees, teas, olives, peanuts, cotton, sunflowers, pineapples, millet, sugar peaches, bamboo, and oranges. Most of us, with a little thought, can add more plants to this list—but how many more?

If we look at J. C. T. Uphof's amazing *Dictionary of Economic Plants* and delve into just one page at random, say page 95, we find among the entries: *Calamus ovoideus,* a palm from Sri Lanka whose young leaves are edible, either raw or cooked; *Calamus rotang* from Bengal and other parts of India, whose stems are made into rattan furniture, ropes, and baskets; *Calandrinia balonensis* from Australia, which has fruit eaten by both indigenous and settler Australians; and *Calanthe mexicana,* an orchid from Central America and the West Indies whose ground petals are effective in stopping nosebleeds! Uphof's book contains more than five hundred pages and describes another ten thousand unfamiliar plant species known to be useful to at least one human culture. The dictionary lays out a veritable treasure trove of foods, beverages, medicines, fibers, dyes, and construction materials far greater than most of us imagine.

Animal diversity has also been a major resource for human cultures worldwide, providing cattle, sheep, pigs, horses, chickens, ducks, turkeys, llamas, alpacas, guinea pigs, guinea fowl, reindeer, geese, donkeys, goats, camels, yaks, buffalo, and a range of fish from sardines to sharks. Animal diversity has also provided more modest but highly valued resources such as honey bees, scallops, oysters, lobsters, and crayfish; and we should not forget that cultures around the world rely on a huge variety of animal foods, including insects such as caterpillars, grasshoppers, ants, and beetle larvae. None of these are eaten out of desperation, but rather because they are abundant, tasty, and nutritious. Fungi have been another important source of nutrition, either through yeasts used in baking and brewing or through edible mushrooms.

Human beings have already made use of many thousands of species. In this book we shall show that, one way or another, our dependence on biological diversity is still growing rapidly. Our exploration of the species with which we share the planet is full of surprises as we find that many species on which we depend are unexpected: new antibiotics from ants and termites; life-saving medicines from leeches and parasitic worms; construction materials from snails and spiders; robots inspired by insects and earthworms; bacteria, fungi, and mites running major industries and public services. Most people believe that our future depends on electronics, computers, and space-age transportation. In fact, that future is equally bound to a host of mostly tiny organisms that we are only beginning to understand. We will see, for example, that the creatures that live in our soils are worth far more to us than all the products of Silicon Valley.

The species of Earth are our biological wealth and, like any capital, should not be squandered or thrown away. As this wealth is revealed in the following pages, it is prudent to assess whether or not we are taking care of it. If we imagine that the ten to twenty million species on Earth are the equivalent of that many safe deposit boxes in a vast bank called Nature, then most of them remain unopened and we are ignorant of the contents. As we open more and more of them, the contents will prove to be of immense value to society. More sobering is the thought that large numbers of these boxes are being destroyed before we know the treasures within.

We share the world with millions of other species. Most of them are tiny and little known to us. What are their names and what do they do? A new branch of science that studies biological diversity is determining the answers to those questions. Innovative technologies have shown us the enormous and previously unsuspected variety of the biological world. Exploring the natural world is a story of the daily discovery of new species and inventive ways of making a living, and this inquiry is the focus of Chapter 2.

In Chapters 3 through 7 we focus on the biological diversity that sustains the natural ecological systems that keep us alive. In addition, we see that in some cases we can build on this diversity to create new industries. Millions of species interact with their environments and with one another, engineering the mechanisms that regulate the air, water, and soils on which we depend. They have evolved technologies, most often at the molecular level, that break down domestic, agricultural, and industrial wastes. They provide the majority of pest control in agriculture, horticulture, and forestry. All these activities are the basis of human civilization. We do not claim for one moment that these systems have developed specifically for human use; rather, they are the natural products of the activities of vast numbers of species that have evolved over millions of years and allowed human beings to create the diversity of cultures we know today.

The kaleidoscope of populations, species and communities that we know as biological diversity generates, among other things, our food supply and therefore is the foundation of all our food industries—farming, ranching, and fisheries, plus of course the businesses such as trucking and machinery manufacture as well as the financial institutions associated with them. Forestry is another industry based on biodiversity—not just the trees but the minute organisms that generate soil fertility and pest control. Tourism, by taking advantage of immense natural collections of species such as those found in national parks and marine reserves, is dependent on biological diversity too. In the last ten years, tourism has become the single largest global industry. All of this economic activity, involving trillions of dollars

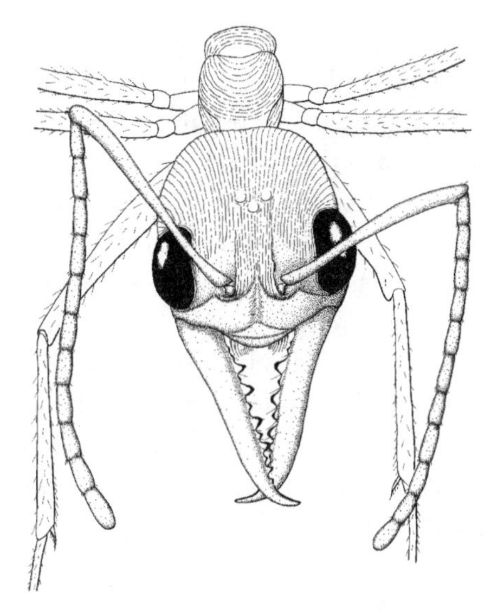

Fig. 1 The front end of an Australian bull ant, a fierce predator that often stumbles upon unwary picnickers. It is also the subject of a patent for a new antibiotic.

annually, starts with the biological diversity that operates the natural systems in which we build our cities and towns.

In Chapters 8 through 12 we explore biological diversity as a source of completely new resources. The two basic requirements are human imagination and the diversity of life on Earth. We introduce a new way to explore biological diversity that takes advantage of everything we know about natural history, ecology, and evolution. Its basis is very simple. If we are looking for a solution to a human problem, we ask an elementary but powerful question: Where might that solution have evolved in the wild? This concept is introduced in Chapter 8, and the following chapters show how a wide variety of species and natural products are rapidly expanding the horizons of established industries and professions such as medicine and pharmaceuticals, pest control, farming, and construction. If the biological resources of the world are represented by a giant iceberg, we have only seen the very tip.

The exploration of biological diversity leads to discoveries that make our way of life more sustainable; that is, they allow us to lead full and productive lives without jeopardizing the resources and the environments that our children, and our children's children, will need. We are then better able to understand our role in the natural world and know how to avoid the big mistakes such as pollution and soil loss. Biological diversity keeps us alive and, beyond that, is the key to a dazzling variety of economic opportunities—many based on obscure, tiny, and often very strange creatures. It is vital that species not go extinct: there is no way of knowing which will turn out to be important (see Figure 1).

We end this book with a chapter about the future. Most people think it wrong to squander wealth or throw it away. Biological wealth is no exception, and conservation efforts are under way in national parks, marine reserves, zoos, botanical gardens, and seed banks around the globe. Industries that are directly dependent on natural ecosystems (farming, grazing, forestry, fisheries, tourism) are searching for ways to preserve the biological wealth, the biological capital, they use. Similarly, human societies, all of which rely on natural ecosystems to dispose of their domestic, industrial, and agricultural wastes, are looking for ways to keep those services intact. To allow extinctions of populations and species is both foolish and selfish—foolish because of the lost opportunities, selfish because of the loss to the next generation, if not to ours.

Why are we so concerned about populations? Simply because populations are where innovation within a species takes place. For example, studies show that a species of grass common in parts of Wales has local populations that are adapted to living on soils containing levels of heavy metals that are normally toxic. This innovation means that the species can occupy a wider range of habitats than most of its competitors. Genes useful for breeding new crop plants often occur in only a few populations of their wild relatives. If those populations are destroyed, the opportunity is gone. Those of us who live in cities have become aware of populations of plant species adapted to colonizing roadsides and railroad tracks, and even populations of birds of prey that have adapted to rooftop nesting and hunting rats and pigeons. All species are made up of many populations, each of which is a little different from the others. Reduction in the number of populations in a species means that it is losing its ability to innovate, to change. It is a danger signal. When it occurs, the species is called threatened or endangered. Many of our most vulnerable species have only a few populations left, and some have only one.

In light of our dependence on biological diversity, it is worth thinking about supply. How much is available and if we mess up, where can we get more? We are often careless with things because we know that they can be replaced—or we believe that modern technology will find a source. Instead, with regard to biological diversity, we should ask: Can we *make* new species, and if not, is there somewhere else we can get them? In spite of the hype that surrounds biological technology, we are still a long way from being able to create life, let alone make a complex organism such as a bacterium or a moose. The scientific reality is that we will not be able to achieve this goal in the foreseeable future. So there is little chance of being able to make new species.

What about finding them elsewhere? No life has yet been found anywhere else in the universe. And if we did find life on some distant planet, would we dare to bring it home? As far as we know, we and all the other species on Earth are alone. *There is no alternative supply;* the organisms with which we live and on which we depend are completely irreplaceable. We cannot create them or obtain them from another source. This stark situation means that exploring, discovering, and understanding the biological diversity of our world is just about the most important task confronting us.

While our approach in this volume is pragmatic, we share with most people the simple feeling that the creatures of Earth are absolutely critical to the human spirit. The fact that all over the world people keep pets, cultivate gardens, maintain aquariums, go bird watching, enjoy nature videos, and take vacations in places of natural beauty, reminds us of another suite of values of biodiversity: those that give us peace and contentment.

Antonie van Leeuwenhoek, who lived from 1632 to 1723, was one of our earliest microscopists. His were exciting times, because microscopes were revealing structures and organisms never before imagined. In 1708 van Leeuwenhoek wrote the sentence at the start of this chapter. Those words remain as true today as they were three centuries ago.

TWO

Exploring a Little-Known Planet

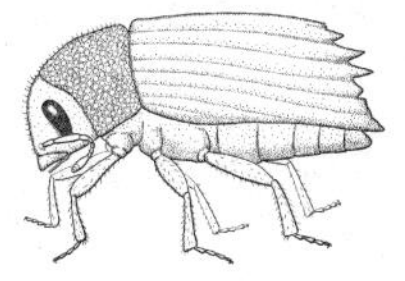

The majority of species on Earth have yet to be discovered. This statement may come as a surprise, since nature documentaries, zoos, and museums present us with a bewildering variety of animals and plants. However, it is generally agreed that right now we know a mere 10 to 20 percent of the species that share the world with us. To achieve some perspective, imagine that all the species in the world are scattered along the 3,900 kilometers of the Mississippi River and that we have embarked on a voyage to discover them. If we start on the delta, close to the city of New Orleans, our current knowledge means that we are plodding along somewhere in Arkansas. The headwaters are still 3,000 kilometers away in northern Minnesota. In this chapter we assert that discovering Earth's new species is the most exciting activity in biology today, and that ours is still a little-known planet.

Every day around the world, biologists specially trained to search out and describe new species unveil organisms previously unknown to science. For example, a recent edition of the *Australian Journal of Entomology* contains descriptions of six new species: a scorpion fly and a mayfly from Tasmania, a bug that feeds on mistletoe, and three mite species—one of which lives only in the feathers of brush turkeys. Species are found in exotic-sounding places such as rain forests, hot springs, polar regions, and ocean depths, but they are also found in grasslands, rivers, lakes, wetlands, and even in back yards.

One of the most sensational discoveries was made in 1994 just 150 kilometers northwest of Sydney, Australia's largest city. The terrain in this area of the state of New South Wales is extremely rugged sandstone ridges and canyons, many of the latter being just a few meters wide but hundreds of meters deep. A team of biologists was exploring a deep canyon in Wollemi National Park when they encountered a strange-looking tree with leaves that resembled those seen only in fossils of the Jurassic age, belonging to species believed to have been extinct for at least 60 million years. In light of Steven Spielberg's *Jurassic Park* movie, the scientists could be forgiven if the hair on the back of their necks stood on end. "Unbelievable" would be a word that truly made sense at this moment: here was a genuine survivor from the age of *Tyrannosaurus rex,* not only alive and well, but, as the scientists looked around, clearly part of a small but healthy colony.

The tree was named *Wollemia nobilis* after its discoverer, David Noble, and because it was found in Wollemi, which is an aboriginal word meaning "watch out" or "look around." Forty adult plants were found, the tallest being 35 meters with a trunk 1 meter in diameter. The tree does not lose its leaves in the usual way, but sheds its lower branches—which fall to create a distinctive litter on the ground. This unusual clutter of dead branches and leaves (it acquired the name Jurassic Bark) was one of the first signs that something unusual was happening in this canyon. The impact of the discovery dramatically reinforced scientific predictions that the world had a great many more species than was once thought. After all, if a new species such as the Wollemi pine, which is taller than any dinosaur, was found for the first time as recently as 1994, not far from a major city, how many smaller species were waiting in the wings? Here was spectacular and concrete evidence that the quest to locate all the species in the world is still in its infancy.

The unearthing of the Wollemi pine attracted media attention around the world, and it was soon known as the dinosaur plant. Yet, it was only the latest of many recent finds of immense significance. In purely scientific terms, the discovery of a new phylum, a new life form, has even greater significance. A phylum is a major group of organisms so distinct from all others that the category is only one step below that of kingdom. Everyone knows that kingdoms (plants, animals, fungi, bacteria) are very different from one another. The phyla are right behind, so that the animal kingdom, for example, contains separate phyla for sponges, mollusks, sea urchins, earthworms, and insects because each of these groups has a basic body plan or structure that is obviously very different from the others. While a discovery of this immensity may seem unlikely, the fact is that two new phyla have been uncovered in the last two decades. The first of these, the Loricifera, was described in 1983. An even more recent discovery, the Cycliophora, is being proposed as a new phylum as we write this book.

Loricifera are microscopic, found in marine sands and gravels and adapted to living in the tiny spaces between the grains. One end consists of a mouth on the end of a tube, the base of which is surrounded by spines; this merges into a pear-shaped body, armored with protective plates. When the animal is disturbed, the entire structure can be retracted inside the spines, which in turn fold like the ribs of an umbrella and disappear into the armored body cavity. While the phylum is widespread, no one knows what it eats or how many species it comprises—although the estimate is that there are more than a hundred.

Cycliophora are also tiny and have an equally unique habitat, the mouthparts of lobsters. The anatomy is baffling, as it has features that belong to several other phyla. Less than a millimeter long, the cycliophoran is basically a barrel-shaped animal with a sticky disc at one end that attaches to the lobster. At the other end, minuscule hairs waft food into the mouth, which leads to a U-shaped gut that empties through an anus uncomfortably close to the mouth. A small lump on one side of the female barrel is the male! Very little is known about this group, but among all its strange features, one aspect was of special interest to those looking for new kinds of organisms. The lobsters on which cycliophora are found are not rare, tropical, or exotic, but a common variety that has been taken for centuries from the sea surrounding northern Europe—right in the back yard of Denmark and Sweden.

Our knowledge of the world's species is so rudimentary that biologists are still arguing over the number of kingdoms! Many believe there are five: the plants, the animals, the fungi, the bacteria, and a fifth—the Protoctista—that contains all the remaining organisms that do not fit into the other four. Protoctists include, for example, the four hundred to five hundred species of slime mold that inhabit rotting leaves and logs (Figure 2). Under the microscope they resemble small pools of jelly that pulsate gently as they glide in search of prey, usually bacteria. They thrive best in moist conditions, but when their environment begins to dry out, the individual blobs of slime come together and construct a small stalk that bears tough, weather-resistant spores at the top, like a bunch of balloons on a stick. While slime molds are odd, they are very common in gardens. Another type of protoctist is the slime net, which is basically a family of cells inhabiting a loose network of slimy threads. The cells move, but only within the threads, which are therefore known as slimeways. Again, these organisms are weird but common; their favorite habitat is the surface of water plants. Still other species are called Problematica, because their bodies are so strange. One, with the formal name *Buddenbrockia plumatellae,* lives inside the minute animals that form mosslike "animal mats" on rocks in tide pools. It is a worm about 3 millimeters long, with four groups of muscles stretching from one end to the other, but it has no digestive, excretory, or nervous system!

While new phyla have recently been discovered and many organisms are so bizarre that they have to be loaded into a new kingdom, microbiological methods currently under development are revealing a previously unsuspected world of bacteria and fungi. In the past, microorganisms have been studied by culturing them on plates of a nutritious jelly called agar. Most of us have seen images of the scientist at the lab bench, complete with flickering bunsen burner, examining

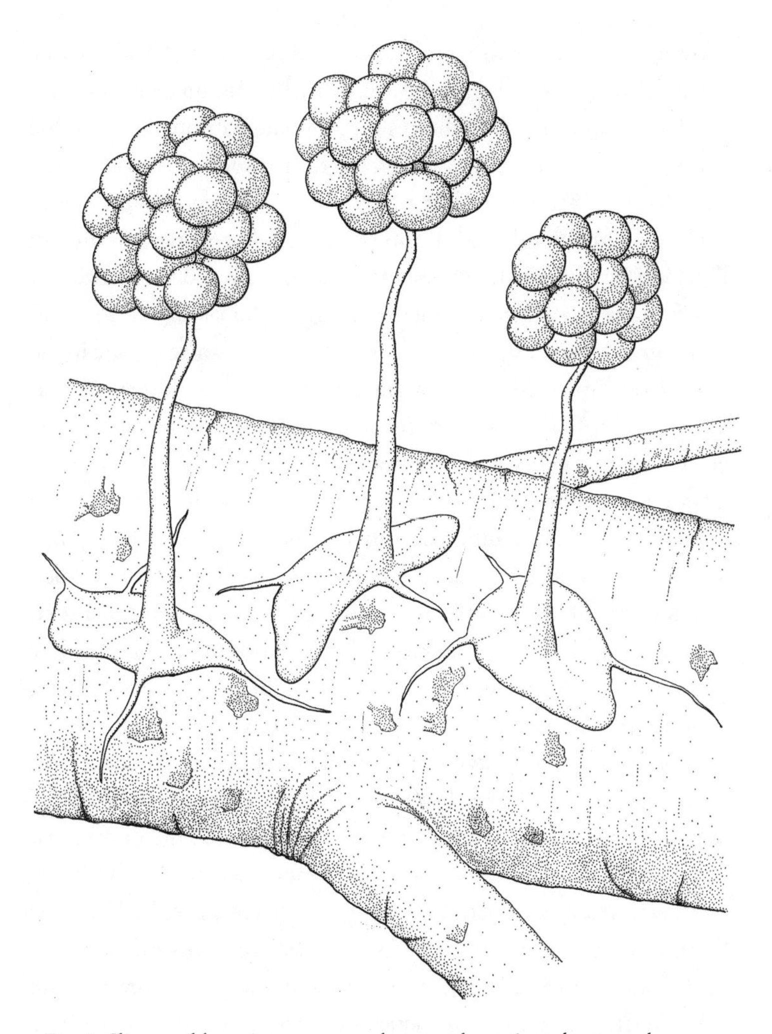

Fig. 2 Slime molds are important predators on bacteria and are seen here on a decaying twig. Individual cells have come together to form tiny communities from which spore-bearing stalks emerge.

colonies of bacteria or fungi in a small glass petri dish. The drawback is that the microbes can only grow if that jelly contains the food they like. The difficulty lies in finding the appropriate nutrition for microbes whose diet we fail to understand. In fact, the problem is so daunting that it is estimated that only 5 percent or less of microbes have been cultured or *can* be cultured. If this is so, the other 95 percent are out there, lurking unknown in nooks and crannies around the world.

Fortunately, molecular methods have revolutionized the field. Now, rather than transferring a swab of bacteria to an agar plate in the hope that the microbes will grow, the scientist extracts the genetic material, the DNA, directly. Current technology makes it possible to determine very quickly whether or not the bearer is new to science. To illustrate the power of these new methods, consider soil samples that were investigated by traditional methods, the bacteria and fungi grown on agar plates. DNA extracted from the same soil samples has revealed hundreds of other microbes, none of which resemble the cultured ones. Thus, not only are the molecular methods powerful, but they are telling us about a vast, previously unimagined diversity of bacteria and fungi that science is only beginning to glimpse.

Some microbiologists conclude from the application of new molecular methods that instead of five kingdoms, there are three superkingdoms or domains. The first two, the Archaea and the Eubacteria, contain only bacteria that, from the microbiologist's point of view, are as different from one another as gorillas are from daffodils. The third domain, the Eukaryota, encompasses everything else—including animals, plants, and fungi because, relatively speaking, those are all more closely related than the two bacterial domains. Whether or not this view stands the test of time, it shows how little we appreciate the true diversity of bacteria.

We have yet to mention viruses, which cannot lead independent lives but in order to reproduce must infect other living cells. Some scientists therefore do not consider them living, but viruses are serious micropredators that attack and kill all sorts of bacteria and other kinds of cells. Since bacteria are food for viruses, they have the potential to affect entire food chains. Thus, what happens at the level of microbes may affect what happens at the level of landscapes! If we include microbial diversity in our analogy of exploring the species of the Mississippi River, we have not even left the suburbs of New Orleans!

To cope with all this uncertainty, most biologists follow the concept of the five kingdoms, each of which contains a variety of phyla. At present the total number of phyla is ninety-six. Put another way, ninety-six very different basic life forms are currently known to inhabit Earth. The full magnitude of this diversity is illustrated by the phylum to which we belong, the Craniata (often known as vertebrates). It may come as a surprise that fish, frogs, toads, lizards, snakes, birds, and mammals all belong to this single phylum because they share the same basic body plan: a vertebral column (backbone) with a brain enclosed by a protective cranium or brain box. The covering of scales, feathers, or fur is relatively unimportant. Thus, nearly all the animals we know as pets or as farm and zoo animals, including goldfish, salmon, bullfrogs, snakes, lizards, parrots, turkeys, eagles, dogs, cows, whales, kangaroos, leopards, elephants—together with the dinosaurs—belong (or belonged) to one single phylum. There are ninety-five others.

To avoid the impression that, other than Wollemi pine, all the newly described species are obscure little animals or bacteria, let us take a brief look at some that have been discovered and described since 1990. In the state of Queensland, Australia, there are five new species of rain-forest lizards, a large tree frog, and several fish species. The fish were discovered in what may sound like an unlikely place—the heart of the arid outback. In a few lonely spots, water seeps from the rocks beneath to form permanent shallow pools and there, quietly over a few million years, unique species of fish have evolved and thrived. A new species of whale was an astonishing recent discovery off the coast of Peru, and this followed the discovery of two new porpoise species. Two new monkey species, one from the southeast reaches of the Amazon and a marmoset from the central Amazon, have been identified in Brazil. Another mammal in the shape of a large, nocturnal rodent has been described for the first time in the Philippines. The local people know all about it, calling it the Cloud-Runner because it inhabits the trees in misty rain forests. A new species of striped rabbit was discovered in Laos when it triggered an automatic camera in the forest; once alerted to its presence, zoologists discovered skins in a local market and made the identification through DNA analysis. Among the hundreds of new species of invertebrate animals is a squid from Hawaii and a giant crayfish from New South Wales. Finally, demonstrating that even bacteria can be big, the sulfur pearl bacterium is the world's largest. It has just been discovered in the mud off the Skeleton Coast of Namibia and is large enough to be seen without a microscope or hand lens. These are just a few of the new species that have come to our attention; the biological literature documents thousands more.

The discovery of many species has been unexpected, not because the discoverer stumbles across them suddenly in the canopy of a rain forest or on some remote island, but because they appear in well-studied collections in the museum or laboratory. Until recently, new species have been distinguished from others only by characteristics that can be seen with the naked eye or under the microscope in preserved specimens. For example, a new plant species may differ from its relatives by the shape or hairiness of the petals; a new bee species may be distinguished from all others because its head has a particular array of bristles. If specimens looked the same, or nearly so, they were judged to belong to the same species. The problem is that the organisms themselves do not necessarily distinguish one another by the features we use, so specimens that look alike to us may turn out to belong to very different species. Thus, when specimens in museums and laboratories have been reexamined by different methods, hundreds of new species have emerged from established collections.

The use of sound is one of these alternative methods. For example, crickets, cicadas, frogs, and bats have been recorded as they call to each other. The calls have been analyzed in the lab, and those of individuals thought to be of the same species (conspecific) were found to differ, often in subtle but distinctive ways. In some cases, playing back the recordings selectively attracted particular animals, supporting the conclusion that they were different species. Thus, sorting by sound has been critical evidence of the presence of different species very similar in appearance.

The extraction and comparison of DNA from various organisms have played a vital role in revealing hidden or cryptic species. These powerful methods have revealed new species of mammals, birds, lizards, fish, and many unknown insects, crustaceans, mites, and worms. Frequently DNA is the only way to isolate new species of bacteria and fungi. One particularly dramatic demonstration involves the velvet worms of Australia. These animals are living fossils, halfway between worms and insects. Based on external, visible characteristics, there were thought to be nine species. However, in the last two decades over a hundred new species have been identified by means of molecular methods. Interestingly, in many cases where scientists have reexamined the specimens identified as new species, they have seen distinctive features that had previously been overlooked.

Improved technologies have changed our perspective on the size of organisms as well. Most people regard the whale as the standard for large size. However, the nemertean or ribbon worm, *Lineus longissimus,* can grow to 30 meters, longer than everything but the blue whale. As for the heaviest organism, rivals include the blue whale at 180,000 kilograms and the giant sequoia, which weighs in at 2,000,000 kilograms. Now there is another nominee: a quaking aspen clone, a mass of trees all sharing the same root system, that covers 43 hectares and is estimated to weigh 6,000,000 kilograms. While these are the true heavyweights, an armillaria fungus has attracted attention, as its body of microscopic threads is estimated to have penetrated 15 hectares of soil and to have a weight calculated at roughly 10,000 kilograms—not bad for a "micro" organism.

These continuous discoveries lead us to the surprising conclusion that the actual number of species in the world is unknown. While it is generally agreed that about one and a half million have been formally named and at least briefly described, at least another five million to ten million exist. Most of the species we know are insects, and this group undoubtedly contains several million more. Other groups may be as large, or larger: the bacteria, the fungi (perhaps a million species), the nematodes or roundworms, and last but not least those tiny eight-legged relatives of spiders, the mites, a group that may harbor another million species. The authors once had the opportunity to look at a handful of forest soil under the microscope and each of us did a classic double take as we realized that about half of the "particles" we saw were slowly moving and turned out to be soil mites. Other relatively little-known groups that may also contain a huge number of species are the single-celled plants known as algae and the single-celled animals, or protozoa. One consequence of our age of continuing discovery is that we can look at any landscape, or seascape, in a new way. We now realize that the world is dominated by small organisms, as shown in Figure 3. The number of species in a group is represented by its size so that the most diverse groups in the world are the bacteria (on the right), then the insects and their relatives. Next are the flowering plants such as trees, then (perhaps surprisingly) the crustaceans. Although they are best known as crabs and lobsters, thousands of mostly tiny crustacean species inhabit the rivers, lakes, and oceans of the world. Then come the fungi, the mollusks, single-celled animals, fish, worms, and finally, at the least diverse (left) end of the scale, all the familiar craniates such as the birds, amphibians, reptiles, and mammals.

Any part of the world is alive with many different species, although special technology may be needed to reveal them. For example, the left half of Figure 4 shows an Australian landscape with eucalyptus trees and a variety of other Aussie plants and animals. The right half illustrates how technology reveals the biodiversity of that scene. At the top is a representation of the species that can be seen by eye, things such as flowering plants, frogs, snails, and caterpillars. In a scene like this it would be possible to count perhaps a hundred species. Suppose the landscape could be examined through a powerful binocular microscope, magnifying everything fifty times. Then one or two thousand different species would be revealed, including mites, termites, insect larvae, and tiny algae and lichens. Magnify the scene five hundred times with an electron microscope, and an entirely different menagerie of microscopic species such as bacteria, fungi, slime molds, and algae appears. A conservative number would be three thousand. It becomes clear that any piece of land or sea harbors not merely the species that immediately catch the eye, but thousands more. The situation is not unlike astronomy; most of the stars in the sky cannot be seen with the naked eye, but with the help of sophisticated technology, millions of stars appear in the cosmos. In biology, it is microscopes that reveal the otherwise invisible objects. In this case they are species and the 'cosmos' is the Earth's myriad environments and habitats.

An important reason why there are so many species is that each one is home to others. Look at an individual tree, say an oak or a gum. It will host a variety of birds, mammals, lizards, insects, worms, and fungi. Some are merely using the tree because it is handy, whereas others are fine-tuned to live on or in that species and that species alone. In other words, the organisms are adapted to that tree species. Thus, many kinds of birds may nest among its leafy branches or in holes in its trunk, but they could just as well nest in any of a dozen other tree species. By contrast, many insects have larvae adapted to eat only the leaves of that tree, or bore only into its trunk, so that if the tree species goes extinct, the insects do too. All species are home to others in this crucial way, as the bark beetle illustrates (Figure 5). This particular species is in fact a community that houses three species of mites, four species of roundworms, three species of fungi—and at least seven species of bacteria live either on or inside the animal. Most need the beetle to survive because it is their only home. Similarly, a study of just one species of fish revealed that it was inhabited by five species of crustaceans and up to nine species of roundworms, tapeworms, and flukes. And a single species of palm in Thailand proved to be home to ninety-five different species of fungi.

Many of the animals that live on or in others are harmful in various way. Caterpillars, for example, eat leaves and are therefore herbivores. In some cases the loss of leaves is negligible, while in other in-

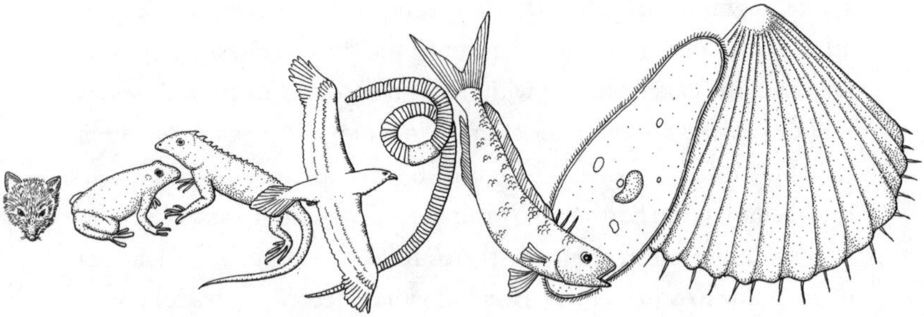

Fig. 3 Which are the most diverse organisms on the planet? Here, estimated diversity is represented by size, ranging from the predominant bacteria on the right to the mammals on the left.

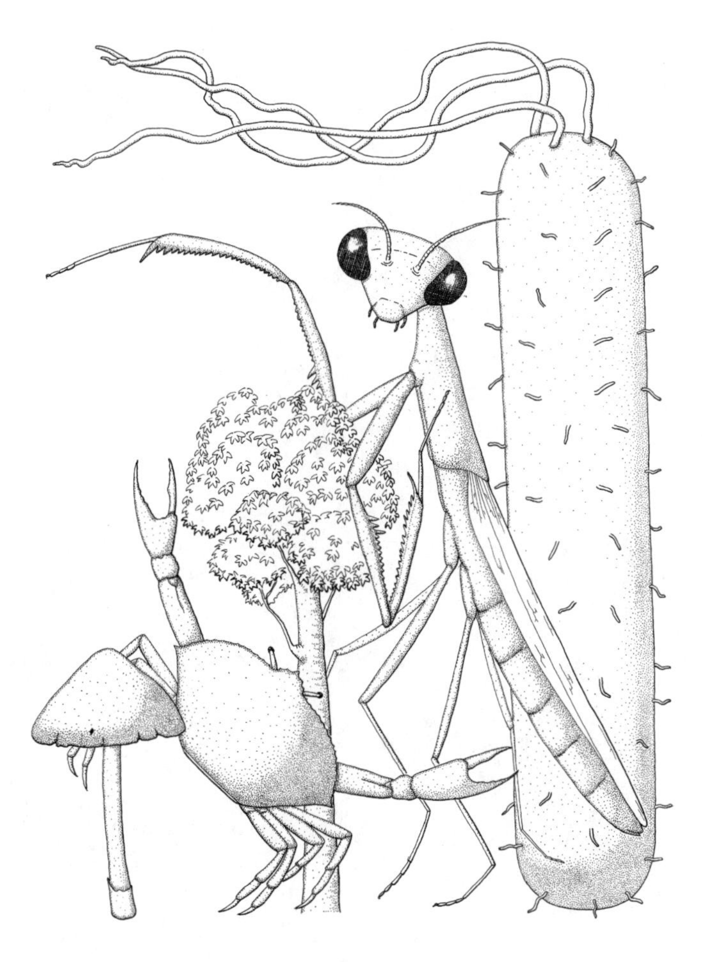

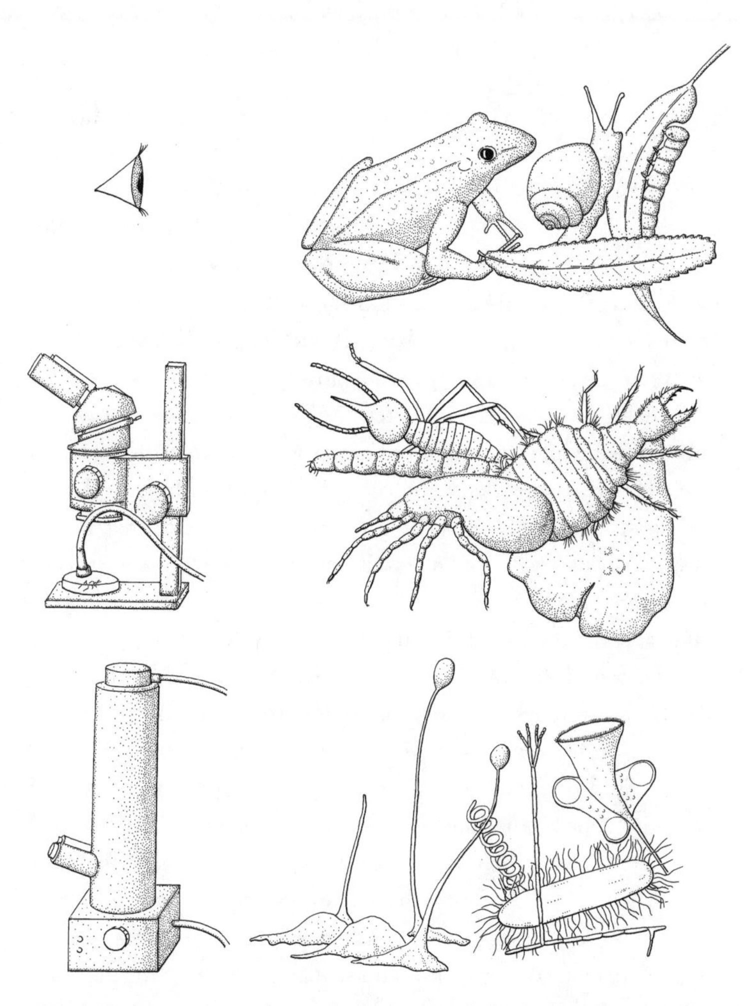

Fig. 4 How many species live in the landscape at left? The unaided eye (right-top), viewing this Australian forest with its gum trees, tree ferns, and termite mounds, could identify approximately 100 species, including frogs, snails, caterpillars, and various plants. If the scene was magnified 50–100 times (right-middle), many smaller species would become visible, including fly larvae, mites, ant lions, and lichens—most likely 1,000–3,000 species in all. Still-higher magnification under an electron microscope (right-bottom) would reveal yet another 3,000–5,000 species, including single-celled animals and plants, fungi, and bacteria.

stances the animals may eat them all, leaving only bare branches; the tree then will not flower that season. Roundworms are often parasites, living in the guts of other animals such as the bark beetle, where they laze around in the stream of predigested food, absorbing what they need through their thin skins. By contrast, mutualists (see Figure 6) live on other organisms and develop a relationship that benefits both. The many different species of bacteria and single-celled animals that live in termite guts are examples. The inside of a termite is a very safe and equable place to live, so the termite provides housing for the microbes. Meanwhile, the microbes digest wood so that the termite is provided with an apparently inexhaustible supply of food. Other examples are the highly specialized fungi that penetrate the roots of trees. The fungi are housed in a well-regulated environment provided by the plant and at the same time provide special threads and enzymes that enable the roots to absorb nutrients from the soil.

The complexity of these associations is illustrated by the prayer plant, a rain-forest species of tropical Central America. The flowers emerge from between small green leaves that secrete tiny, glistening drops of fluid. This nectar consists of sugars and other tasty treats that are avidly collected by ants. Because the ants regard the nectar as theirs, they defend their leafy territory against all comers, including insects such as beetles and moths that attempt to lay their eggs on the plant. In preventing egg laying and hence invasion by hungry larvae, the ants are de facto ant-guards and reduce the damage the plant sustains from insect herbivores. But the story does not end here. Quite often one finds a plant that has a platoon of ant-guards patrolling the flowers, occasionally slipping away to sip nectar, yet clearly is heavily damaged. In spite of the guards, something is eating the flowers. Close inspection reveals that one of the little green leaves between the flowers is not a leaf at all but a slender, flattened caterpillar quietly munching the nearest flower. Its leaflike color and shape might seem adequate to fool the ant-guards, but by appeasing the ants with its own reward, the

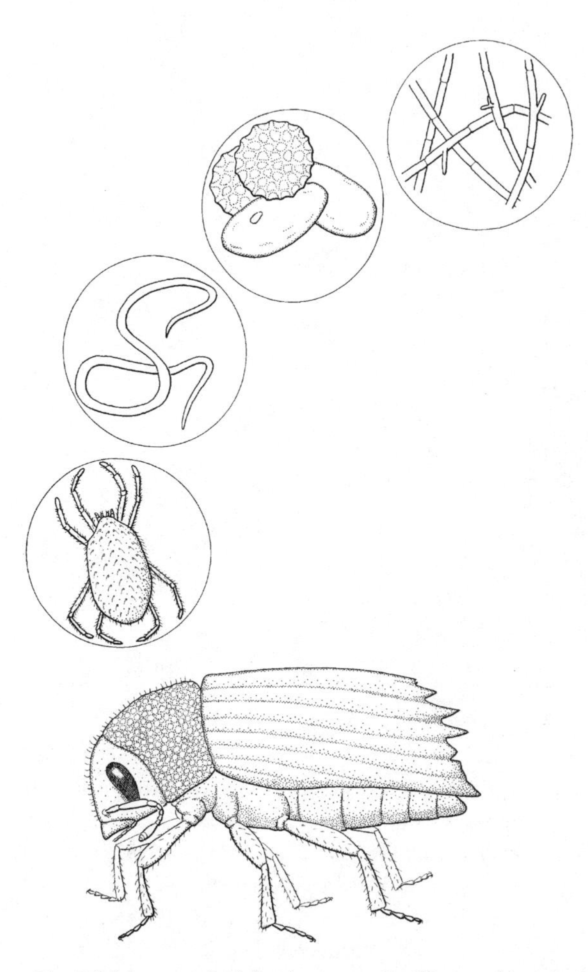

Fig. 5 The average bark beetle, a mere 5 millimeters long, is home to three species of mites, four species of roundworms, seven species of bacteria, and three species of fungi.

Fig. 6 This hummingbird has a close partnership or mutualism with the scarlet gilia plant. The flowers contain nutritious nectar, and while the bird visits different flowers to feed, pollen deposited on its head feathers is transported from plant to plant. The scarlet gilia is a source of food for the bird, and the bird, as pollinator, is an irreplaceable link in the chain leading to the production of seeds.

caterpillar makes absolutely sure that it is not attacked. A nectar-producing nipple on its back end sticks out among the real leaves and the flowers. The ants, completely fooled, sip the caterpillar nectar as though it were coming from a leaf. So here in a single plant on the forest floor is a miniature, highly specialized community of ants together with the herbivores they attack and the herbivores that dupe them.

Is the human body a community too? That depends partly on where you live, how you live, and what you do. A poor farmer in the tropics may suffer from a variety of bacterial diseases—and internal parasites such as tapeworms and flukes, as well as external parasites including ticks and lice. By contrast, a doctor working in a country with long, cold winters may well escape most of these afflictions. Even when we set aside these kinds of differences, our bodies are still crowded with other species.

Antonie van Leeuwenhoek was the first to understand this fact: "For my part I judge from myself that all the people living in our United Netherlands are not as many as the living animals that I carry in my mouth this very day." Indeed, our mouths, noses, throats, stomachs, intestines, and bowels are all home to a diversity of viruses, bacteria, and yeasts. Our skin is home to many bacteria and fungi and often (even in the cleanest house) to head lice and hair-follicle mites. When we asked our colleagues how many different species of bacteria and fungi consider us their home, the estimates ranged from two hundred to three hundred, including *Candida, Staphylococcus, Escherichia coli* (*E. coli*), and *Salmonella.* Our small intestine holds about a million microorganisms per gram, and our bowels about a *billion* per gram. In our bodies, therefore, the human cells are outnumbered about ten to one by the creatures that inhabit us!

Because each species is home to others, it is hard to imagine how many different kinds of homes, or habitats, exist in the world. Various parasites find homes in the lungs of orangutans, the kidneys of squids, or the nasal passages of snakes. Others specialize on the breathing tubes of honey bees, the ears of moths, the nostrils of hummingbirds, the feather shafts of birds, the antennae of ants, the eyebrows of humans, and the digestive tracts of sea urchins. Many single-celled algae live within the tissues of corals, a habit that is extremely beneficial to the corals that are being threatened by the gradual warming of the oceans. A variety of insects and their relatives induce plants to become their homes by disrupting the physiology of a leaf or stem in such a way that a hollow gall develops, into which the insects move or lay their eggs. The inhabitants (which may be moth larvae, mites, or wasps) are then protected by the same tough or even woody tissues with which the host plant protects itself.

When we think of wasps, thousands of species are known as parasitoids—neither predators nor parasites, but something in between. A typical parasitoid wasp (Figure 7) finds a caterpillar or some other larva and sticks an egg on the outside. When it hatches, the tiny larva bores its way into the host and sets about eating it, starting with the least important parts and finishing, when it is fully grown, with the vital organs. It then pupates inside the carcass, emerging some time later as an adult wasp. It is a dramatic scene for anyone lucky enough to witness it! For example, a large butterfly caterpillar infested with dozens of parasitoid larvae over the days gets slower and slower and thinner and thinner. Finally it dies. Within a day or two, the body wall of the cadaver erupts and the caterpillar, instead of turning into a pupa and then a butterfly, produces a small cloud of wasps. There are many thousands of these wasp species each specializing on a different host. A very large number are less than a millimeter in length and easy to miss; most people have never heard of them. However, they are a lifesaver for many farmers, for the simple reason that some crop pests are the prey of parasitoids. By carefully breeding the wasps in special factories and then releasing them in the fields, chemical-free pest control becomes a reality.

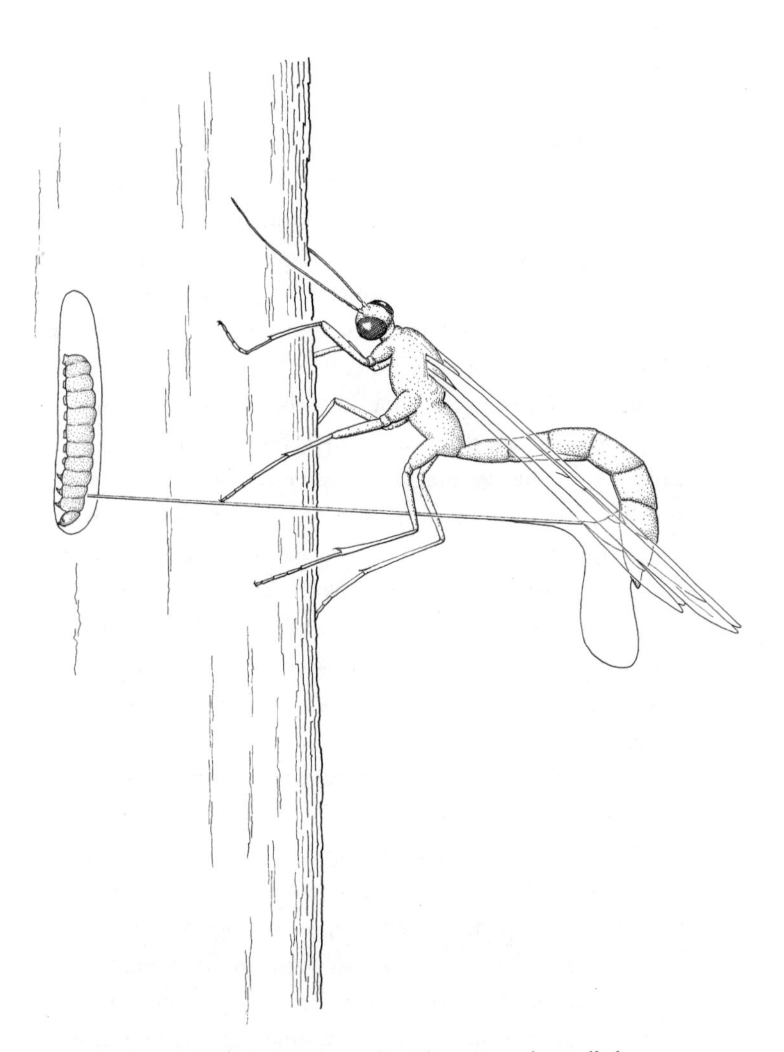

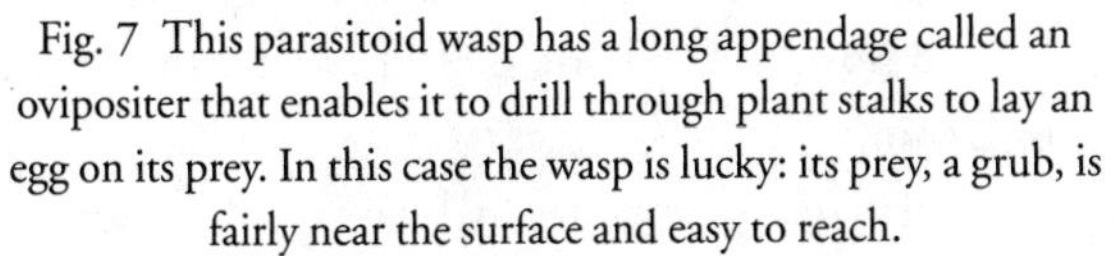

Fig. 7 This parasitoid wasp has a long appendage called an ovipositer that enables it to drill through plant stalks to lay an egg on its prey. In this case the wasp is lucky: its prey, a grub, is fairly near the surface and easy to reach.

The space between sand grains is the habitat of several minute invertebrates, including the loriciferans we have already discussed. Recent exploration has revealed even more remarkable habitats. In Antarctica certain frozen lakes were always thought to be devoid of life. However, some of their dark-colored bottom sediments absorb enough energy from the sun to warm small pockets of ice, resulting in tiny bubbles of water—bubbles teeming with bacteria. Then there is Lake Vostok, which lies 4 kilometers beneath the ice of East Antarctica, is 14,000 square kilometers in area, and is up to 600 meters deep. It has been sealed off from the rest of the world by ice for millions of years, and scientists have just discovered it too is full of bacterial life.

Exploring another seemingly inhospitable environment, microbiologists have looked for bacteria at increasing depths in the Earth's crust. Much to their amazement, they have found abundant bacteria half a kilometer down. In very hard rock the bacteria are confined to cracks and crevices, but rocks formed from sediment still contain plenty of room for microbes. In addition, mini–food chains lie hundreds of meters below the surface, where single-celled predators hunt and devour their bacterial prey.

A further dimension of the rock bacteria is even more fascinating. In sedimentary rocks made from particles of sand or mud accumulated slowly over long periods, the bacteria may have been simply another type of particle laid down. Microbiologists speculate that some subterranean bacteria, having been trapped eons ago as the materials accumulated and the rocks formed, may represent life forms that existed many hundreds of millions of years ago—living fossils. If this is the case, then exploration of rock-inhabiting bacteria that are still alive could reveal secrets of the origin of life on Earth. Such research has only just begun.

Of all the relatively unexplored habitats of the world, the most promising is the ocean that covers nearly three quarters of the Earth's surface. Marine exploration is still in its early phases, but even now an astonishing variety of new organisms is being found. In the open ocean, animals known as salps drift on the ocean currents, conjuring up images of space stations as they aggregate like so many space modules, indifferent to their bulk because of the buoyancy of the water, and to their size because of the vastness of the ocean. Each animal is barrel shaped, several centimeters long, and bound to its neighbors by a tough, elastic material. The colony that develops can be several meters across and, although largely transparent, can suddenly glow with bright luminescence. Contracting muscles that embrace each animal, like hoops on a barrel, squeeze the water through its digestive system, which extracts the microscopic organisms.

Study of marine microorganisms has turned up many species including a new group of bacteria called *Prochlorococcus,* discovered early in the 1990s. The organisms are minute but extremely abundant. Because they can convert the energy of the sun into basic food molecules just as plants do, their vast numbers are just as important to the life of the ocean as forests are to the land—perhaps more so.

These bacteria are one component of plankton, which is a mixture of many kinds of small organisms such as single-celled algae and small crustaceans that are the foundation of oceanic food chains. However, the bacteria have much smaller bodies than these organisms; few analysts had suspected their presence. Thus, the mesh size of the nets used to trawl for planktonic organisms, although tiny, was simply not minute enough to capture bacteria. Smaller mesh sizes captured this ultrasmall plankton (known as picoplankton) that is now known to be very rich in species. Its presence means that even a patch of clear ocean water is, in fact, a thin soup of millions of microscopic organisms. Current methods for the exploration of ocean picoplankton involve extracting DNA directly from seawater.

Sometimes the DNA revealed is so different from that which is known that scientists wonder if they have come across new phyla. This expectation is not surprising because the oceans already contain more animal phyla than any other environment. Twenty-eight of thirty-seven animal phyla exist in the oceans, thirteen of which are not found anywhere else on land or in freshwater. In contrast, there are eleven terrestrial phyla, only one of which is confined to the land. This diversity of basic body plans is one reason for believing that the ocean environment has many more secrets to reveal. Analyses of the seas surrounding the United States, for example, have led sober scientists to speculate that between one million and ten million more species lurk beneath the waves. Even if further research shows that the lower estimate is more accurate, our view of the total number of species is changing radically.

Perhaps our appreciation of biological diversity in the sea has lagged because we have been looking for big creatures such as whales, giant squids, sharks, or fish large enough to eat. As the technologies have become available to filter and analyze seawater and dredge up sediment, it has become clear that most marine species are small, minute, or microscopic. We have already touched on the filtering methods that have revealed that open water is often densely inhabited by plankton and picoplankton. Let us now dive to the ocean floor, to examine the complex, thriving communities whose existence was unsuspected even a few years ago.

One of the most dramatic ocean-floor studies took place off the northeast coast of the United States. Samples taken at depths between 1,500 and 2,100 meters (with one at 2,500 meters) were each quite small. The total area covered was a meager 21 square meters, about equal to the floor plan of a small house. The results astounded the scientific community. This environment, formerly thought to be a sort of muddy desert, yielded 798 species, 171 families, and 14 phyla. The most diverse groups were mollusks, worms, crustacea, sea urchins, and their relatives. Based on this result, at least one new species should be found for every square kilometer added to the study.

Although adding new areas is not possible because of limits on the available resources, it is possible to extrapolate with respect to the total number of species in the oceans. If we say that the average depth of this study was 1,800 meters and the atlas tells us that globally the area of ocean at this depth is roughly 300 million square kilometers, then, if there is one new species for every square kilometer at this depth, there should be about 300 million new species. In fact, biologists do not predict anywhere near this number, because some of the assumptions are obviously wrong. For example, the environment at this depth over the whole world is clearly not identical to the one studied. Even so, we perform the calculation to illustrate two points: first, the difficulty of determining accurately how many species live in the ocean (and on Earth, for that matter); and second, the reasons why modern biologists believe many more species await discovery.

We know the ocean floor is not universally the same because in places it belches scalding gases and fluids. These black smokers, or hydrothermal vents, are cauldrons of volcanic activity; thus it is all the more incredible that they are inhabited by an astonishing variety of sea creatures (see Figure 8). First discovered in the mid-seventies along volcanic ridges in the deep oceans, they appear to be characterized by the very factors normally regarded as discouraging life: very high temperatures (up to 350 degrees Centigrade), noxious gases such as ammonia and methane, heavy metals brought up from deep below, and even strong doses of natural radioactivity. Yet these apparently hostile conditions support many kinds of organisms, often in very large numbers. Aside from the abundant and ubiquitous bacteria, there are tube worms up to 2 meters long, clams, mussels, and shrimps. Almost none has been found anywhere else and until a few years ago they were totally unknown to science. The current inventory is 360 species from 200 genera and 65 families.

As if the living conditions on black smokers were not bad enough, the vents regularly erupt, belch, and disappear; organisms that depend on them must therefore cope with the apparent chaos of a constantly

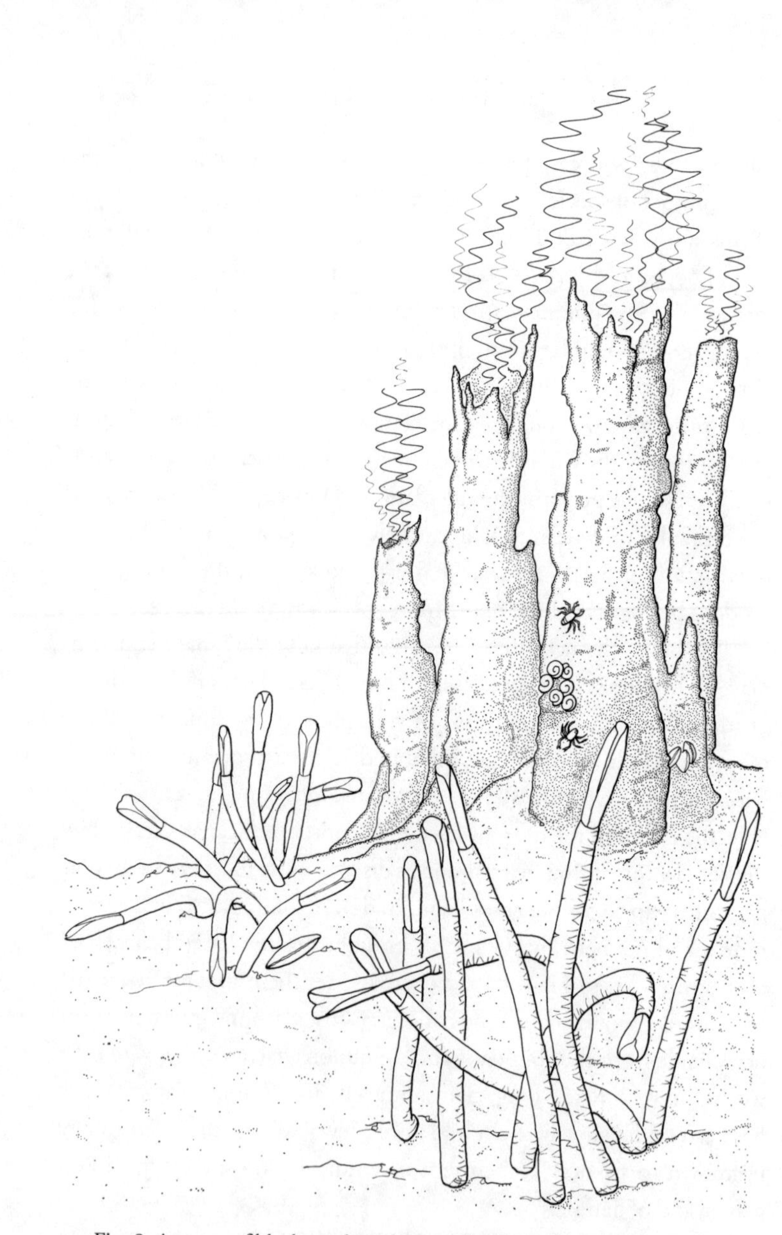

Fig. 8 A group of black smokers deep on the ocean floor. In spite of the high temperatures and poisonous gases, they are home to a variety of tube worms, seen in the foreground, and shrimps and mussels such as those clinging to the smoker itself.

shifting habitat. A great many of these organisms have larvae that appear to be able to travel long distances between vents. However, a chance discovery perhaps gives a clue to their mode of dispersion. Scientists in a deep-sea submersible off the coast of California came across the carcass of a large whale on the ocean floor that was teeming with many seabed species. Some did not come from the ooze on which the whale was lying, but appeared to be species known to inhabit black smokers. Was the huge corpse a relay station, accumulating adults that in turn pumped out more larvae in search of vents?

The many habitats in the ocean continue to surprise us. For example, the coral *Lophelia pertusa* is found in many parts of the world—not in the warm tropics as expected but in cold, deep water such as the North Atlantic off Britain and Norway. The mud on the seafloor in these waters is extremely rich in species, not least the nematodes or roundworms and some truly enigmatic organisms, the Xenophyophora. Because these last resemble both sponges and single-cell animals and fit nowhere else, they are classified as protoctists. Some appear to consist of a single giant cell 10 centimeters across, others as branched tubes of jelly, but nearly always they are wrapped in granular coats of particles from the seabed.

Divers exploring ancient caves and volcanic lava tubes connected to the ocean and filled with seawater have found communities of new species. One in particular stunned the zoological world. At first sight it looked like an unlikely cross between a millipede and a shrimp, swimming upside down in the darkness. In fact, it was the first discovery of a new class of crustaceans that was given the name Remipedes. A class is just below a phylum in the taxonomic hierarchy, so this animal was strikingly different from any other living organism.

Forest canopies come close behind the oceans as places where new species abound. In the tropical rain forests of Central and South America, central Africa, Southeast Asia and northern Australia, biologists are finding improved ways to reach these treetop communities.

Most frequently, biologists fire an arrow over a branch high overhead. When the arrow falls to earth, the slender line attached to it is used to pull a rope up and over the branch. At the end of the rope is an insecticide fogger much like those used to spray mosquitoes. All this activity immediately frightens away the larger animals such as snakes, lizards, birds, and mammals—which the biologists have no desire to kill. Even if they escape, their diversity is negligible compared to what is about to tumble out of the trees. The insecticides are biodegradable pyrethrums that are pumped out among the leaves in great clouds. Special sheets are spread on the ground below, and soon hundreds of stunned or dead insects (and related animals such as spiders and mites) rain to the ground.

Most of the scientists involved in this kind of research are extremely experienced field biologists. Often even they cannot believe their eyes: so many small animals live up in the canopy and so many are completely new to science. The canopies of a small group of trees usually yield tens of thousands of specimens and many hundreds of species. This harvest may take years to identify and classify and probably will involve biologists with different specialties from all over the world.

More recently, it has been realized that while the fogging method can produce vast numbers of specimens and species, its value is limited because the insecticide reveals only the animals on the canopy surfaces. It cannot reach the insects and other small animals that burrow into flower or leaf buds, or the many that live in the thin layer between the upper and lower surfaces of leaves, or those that hide under the bark or bore into the wood. It also tells us little about the animal's life style, since its first appearance is as a corpse on the sheet below. To remedy this, biologists have borrowed an idea from construction engineering and installed giant cranes with long booms in the middle of the forest. The buckets that rise above the canopy and travel along the boom carry biologists directly to the particular branch on the particular tree they wish to study. They may then videotape the action or collect from the surface with a net, or from the interior with a saw or drill, to reveal more of the secrets of this previously remote environment.

While some habitats like forest canopies are difficult to explore, others are easy. You can stand on them, often not suspecting what is beneath your feet. For example, the surface of many desert areas is not the lifeless sand or dust it may seem, but rather a crust of organisms adapted to life in this extraordinarily harsh environment. The crust is, in fact, a community of very different organisms, including algae, mosses, lichens, fungi, bacteria, and cyanobacteria. They need water, which they absorb from dew and fog as well as rain, but otherwise the community is remarkably independent. It is able to harness energy from the sun to make food, and can absorb minerals from the soil below and nitrogen from the air above. The species involved are extremely valuable to the desert flora and fauna, in that they cover the surface, cutting down wind erosion and absorbing water and nitrogen. These critical activities are one reason why desert biologists are so interested in apparently lifeless and somewhat ugly crusts, and why conservationists are so concerned when they are broken up by trampling hooves or the crushing wheels of four-wheel-drive recreational vehicles.

All these examples show that Earth's biological diversity is amazingly complex and interdependent. There is every reason to believe that biodiversity will continue to be one of the most exciting subjects of science for many years to come. We have seen that it may require great physical stamina and innovative technologies. New organisms are being found, sometimes in bewildering variety, in places no one thought possible. The number of species, genera, families, and occasionally classes and phyla continues to expand as does appreciation of the variety of life styles and adaptations required to survive in so many different habitats. If any of you doubt this, try counting all the species that live in your garden—from the tallest plants to the microbes. We guarantee that you will still be counting when you get to a thousand, and you, knowingly or otherwise, may have encountered a species new to science.

Now that you are armed with this knowledge, we will introduce the myriad of species that run the ecosystems that keep us alive. And it is in the context of this enormous variety of biological invention and triumph over hardship that we can look profitably, and nondestructively, for help with our own problems. We can explore biological diversity for wild solutions.

THREE
Basic Survival

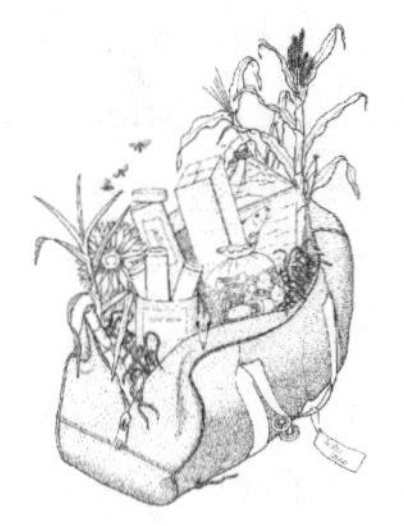

Biodiversity delivers a wide variety of free life support services needed by civilization. These natural services derive from the functioning of ecological systems—ecosystems—that comprise the communities of plants, animals, and microorganisms in an area, and their interactions with one another and with the physical environment. Major ecosystems, such as forests, prairies, lakes, deserts, and oceans have been in place so long and are so ubiquitous that people tend to take for granted the services they provide. However, the lives of the organisms in those ecosystems are firmly intertwined with our own lives.

One vital ecosystem service is maintaining the quality of the atmosphere. It is critical for many reasons in its own right: we would rather have oxygen to breathe than ammonia. It is nice not to live on a planet so hot that the temperature in New York in July is frequently 40 degrees centigrade—which could happen if the grip of ecosystems on the carbon dioxide content of the atmosphere slackened. But control of atmospheric gases is tied to other services as well. One of the most important gases in the atmosphere is water vapor, and its amount and distribution are tightly connected to the way liquid water circulates on our planet—that is, to the hydrologic cycle, which among other things supplies us with fresh water. Forests and other vegetation play a major role in the hydrologic cycle, both pumping water out of the soil into the atmosphere and with their roots holding soil in place, which provides a sponge to soak up heavy rains. This action in turn contributes to another service that is crucial for the huge proportion of humanity that dwells in river valleys—the control of floods.

Many other valuable natural services are supplied to us by ecosystems. The preponderance of pests that attack our crops, or could attack our crops or could carry diseases to us, are kept under control by the natural pest control services of ecosystems. Ecosystems are also responsible for the generation of soil—the grinding up of rocks and the formation of the rich organic materials that contain and nourish microorganisms. These activities recycle the nutrients that are essential to all plants and animals.

An ecosystem service that is less familiar is the maintenance of a natural genetic library. All species contain DNA, a chemical that is the most compact and efficient form of information storage known. Genes are made of DNA, and each one provides some of the information necessary to make, control, and maintain a species. We get an idea of the compactness when we realize that the total weight of DNA required to reproduce the entire human species, about 6 billion people, is about 50 milligrams, the weight of a small match. When we remember that there are millions of different species, it is evident that the DNA on Earth is like a vast genetic library and that each species is a book in that library. Humanity has withdrawn, or may withdraw, a wide variety of useful (often essential) "books" from that library, ranging from the species that were the ancestors of crops and domestic animals to those that supply us with pharmaceuticals. The potential benefits of some books in the genetic library are only beginning to be appreciated: antibiotics could be derived from the toxic substances that ants in damp, earthy tunnels use to keep fungi from growing on them, substances that we shall describe in some detail. One might think of any of these materials as goods, supplied to us as a service of ecosystems. Other ecosystem goods come to us in a continual flow, from fishes that we capture in seas, lakes, and rivers to timber that we harvest from forests.

Forests, especially tropical rain forests, are genetic libraries containing diversity at many different levels: of species, of populations that constitute each species, and of individuals that make up every population. All this genetic diversity provides great flexibility in the way forests respond to change. For example, a forest that has many different tree species, each genetically diversified, is more likely if it is stressed—say, by shifts in the climate—to change in composition and persist, rather than simply disappear.

Perhaps the best way of becoming more familiar with ecosystem services is to use a "thought experiment" originally suggested by John Holdren and developed by Gretchen Daily in her book *Nature's Services.* The thought experiment is rather straightforward. Imagine that you are to command the first expedition to colonize the moon, and that you will leave in a year. One of your most important tasks is to decide what plants, animals, and microorganisms the colonists should take along in order to survive and prosper. To make the thought experiment relatively simple, imagine that the moon already has an atmosphere made up of 21 percent oxygen and 78 percent nitrogen, with an ozone layer and some carbon dioxide—a breatheable atmosphere that filters out poisonous ultraviolet radiation from the sun. Assume also that there is already a lot of ground-up rock that is the physical basis of soil, and that plenty of water is available.

Given that physical setting, what organisms would you want to accompany you? Let's pretend that you have a *very* large spaceship (not a problem in a thought experiment). Probably the most obvious luggage would be organisms that can be used for food. You would probably want more than rice, wheat, and corn—the three grasses that make up humanity's main food base—plus beans and a few other items to provide nutritional balance. In fact, humanity commonly eats hundreds of species of plants; if you enjoy a diversified diet, you would probably take many of them with you—apples, bananas, carrots, dates, eggplants, figs, grapes, and so on. Then, of course, there are the animals people like to eat. So the spaceship would have to be loaded with small groups of cattle, sheep, pigs, chickens, and turkeys, as well as aquariums full of tuna, trout, salmon, shrimp, oysters, and the like. Coffee plants, cacao trees, and tea bushes would need to go along. If you have relatively exotic tastes, you might include deer, buffalo, and maybe some of the tastier African antelopes, pheasants, truffles, mangoes, and other delicacies, to say nothing of cloves, peppers, rosemary, and an array of other spices. With a few hundred different species, you would be reasonably assured of a rich and varied diet (see Figure 9).

If life on the moon is to remind you at all of your home planet, you probably want to take along some plants and animals to supply ingredients other than food. Some tree species that do not yield edible fruit would allow the crafting of comfortable, attractive furniture and relieve the monotony of plastic or aluminum housing (or caves) with wooden floors and decorations. Cotton plants would provide comfortable sheets and clothing. Although it might well be possible to survive amid the plastics and artificial fibers you might take with you, many would consider the resulting quality of life to be less than desirable.

You would also need plants that supply other products, so that you could have the benefit of newspapers, magazines, books, furniture, and many other objects to which you are accustomed. Ornamental plants would take you back to the green hills of Earth. Dogs, cats, aquarium fishes, and so on might help to fight off homesickness. You might want to stock a small zoo for entertainment or esthetic purposes. So a few hundred additional species could go a long way toward providing you with a healthy Earthlike environment.

Or would they? It seems we may have neglected a few items. Indeed, even if the "sports car" collection we have been talking about

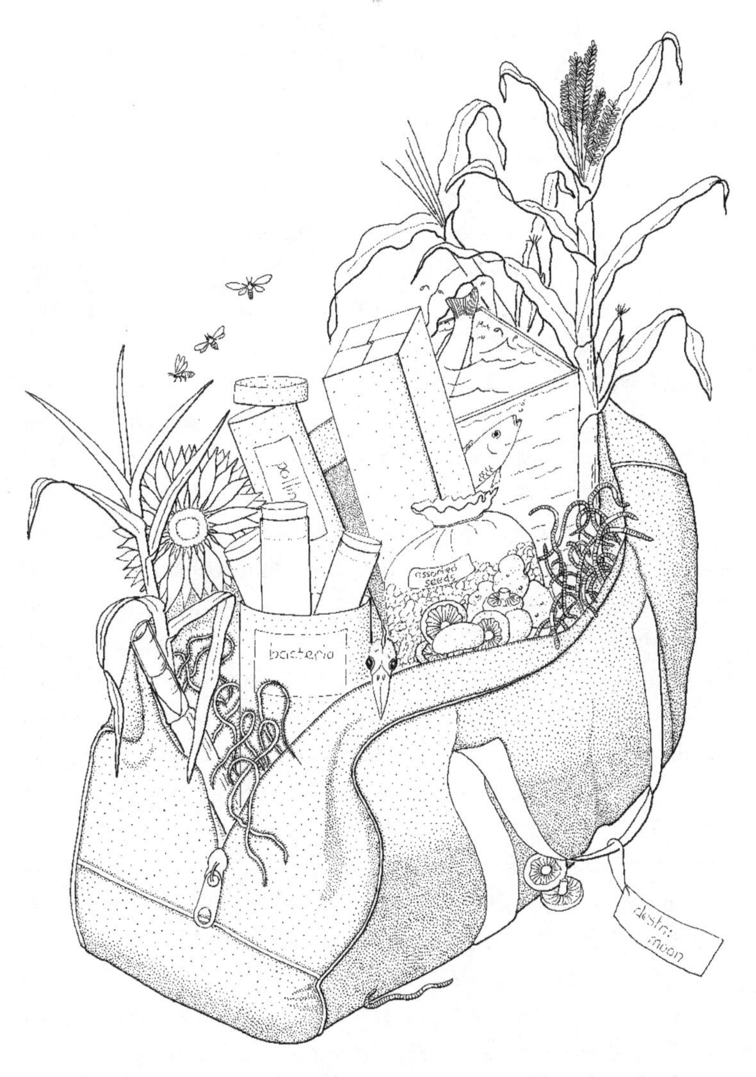

Fig. 9 Moon luggage: a few of the organisms you would need to take along to establish a permanent colony on the moon.

were pared down to the simplest family-sedan, only-for-survival list, a lot would still be missing. What we've forgotten are the creatures responsible for maintaining ecosystem functions. For example, what are your domestic animals going to eat? The expedition could feed some of them on the plants the crew members themselves eat, although it would probably be advisable to bring alfalfa and some grasses, herbs, and shrubs to provide forage for the plant-eating animals.

But how are you going to grow the crops, trees, and other plants? They will not flourish in ground-up rock; they have to have rich soil, and different species may require different kinds of soil. The "richness" of the soil really refers to the presence of appropriate nutrients and of the microorganisms that make those nutrients available to the plants. The microbes occur in vast numbers. For example, a pinch of soil may contain as many as 30,000 protozoa, 50,000 cells of algae, 400,000 fungi, billions of bacteria, and uncounted (literally) viruses. Among other tasks, those microorganisms remove nitrogen from the atmosphere and convert it into forms that vegetation can employ. Other microbes physically transfer vital nutrients from the soil into plant roots. Some trees, for example, will not grow at all without special fungi that perform that function. Further, many different organisms, especially microorganisms, are involved in converting the remains of plants and animals (and their droppings) into those nutrients, which then can be employed again in plant growth.

So a variety of the animals, fungi, and microorganisms involved in the recycling of nutrients would have to be included on your spaceship. Of course, you might have a problem specifying exactly which organisms you need—and in many cases they are very difficult to identify. However, by simply packing enough soil samples from various parts of our planet, you probably could take along most of what you need for soil creation and maintenance.

You still would not be done, though, because about 70 percent of our crops require pollination by various kinds of animals. You would have to consider carefully what crops you are taking, and then which pollinators you would need to ensure that the crops could survive. Honey bees could do some of the work, but when you take honey bees you have to think of whether you will have enough kinds of flowers to supply their hives with energy throughout the season (not just when they are pollinating, say, your apple trees). A further array of pollinator-supporting organisms would need to be included, plants that bloom continually or at staggered times of year, to ensure that the ecosystem service of pollination does not falter.

As you put your plants and domestic animals aboard the spaceship, you would need to realize that you probably could not eliminate all microbial and insect enemies that might be in your organisms, or stowed away with them. You would need to bring natural enemies of the potential pests, since we know that in the long run antibiotics and chemical pesticides will not do the job. Sooner or later, the pests and diseases develop resistance to the chemical defenses, and then they can overwhelm the plants and livestock. Unhappily, determining what natural enemies to take in order to maintain the pest-control service involves a lot of study—to predict what pests might actually make it to the moon, and what predators might be capable of suppressing the pests and maintaining themselves over the long run. You have quite a research project already, and the year until you depart is starting to look like very little time.

Even if you solve all of those problems, you have not yet paid attention to the indirect ways in which wild organisms maintain happy, healthy lives. All you have is the basic raw materials to set up systems to supply the food you will need—a major triumph that is necessary, but not sufficient. The rains, in conjunction with the abundant water we have assumed, will quickly wash away much of your soil unless you have plants chosen as groundcover to break the force of downpours, and with root systems that will hold the soil in place. Indeed, if the moon colony is truly water rich, forests may well be necessary to provide the same type of flood-control and water-purification services as on Earth. (The degradation of the flood-control ecosystem service was dramatically demonstrated to the world in 1998 through the widespread devastation caused by Hurricane Mitch in Central America. Thousands of people died in floods and mudslides, some of which would not have occurred if watersheds had not been denuded of their forests.) But sustaining moon forests opens up the need for another array of pollinator, seed-disperser, and pest-control organisms, complexifying your choices still further.

If your moon colony is expected to operate independently of Earth, you are faced with bringing along multiple strains of each plant and animal species you consider essential. Maintaining that vast genetic library is an absolutely essential ecosystem service on Earth, crucial for keeping humanity in the game of high-yield agriculture. New strains of pests are constantly evolving to attack crops, and plant geneticists must continually try to breed strains of crops that are resistant to these new pests. It is a continuous "coevolutionary race," in which each participant must evolve counterploys against those developed by its opponent. The penalty for losing can be extinction.

Geneticists must have plant strains with different genetic backgrounds (genetic variability) as the basis for developing crops to meet the challenges of new pests and pathogens. The lunar colony would need to be in a position to deal with "surprises" like the 1970 loss of some 17 percent of the anticipated corn crop in the United States. The cause was a newly evolved strain of the fungus causing southern corn-leaf blight, and the disease was countered by using gene resistance from the genetic library. At about the same time, the "grassy stunt" virus threatened the rice crop of Asia. The genes that were required to make commercial rice resistant to the virus were obtained from its wild relatives.

You will have noticed that we put the word "surprises" above in quotation marks. To an evolutionary biologist, the appearance of the blight and the virus were not surprises at all. Wild plants and their pathogens are in a continuous coevolutionary race, the plant evolving ways to resist the pathogen, and the pathogen evolving ways to overcome the resistance mechanisms of the plant. Occasional dramatic victories in such races are to be expected. But crops are generally unable to evolve resistance on their own. As domestic strains, they usually lack the requisite genetic variability and depend on human intervention to acquire new genes for resistance to pathogens or for any other "engineering" advance.

Furthermore, it is quite possible that a lunar human settlement could begin to alter the moon's hypothesized climate just as people may now be altering Earth's climate. Even if the lunar climate were to change only naturally (as Earth's once did), a prudent expedition leader would provide future generations of moon people with the tools necessary to alter their crops genetically to meet the exigencies of such change. As in the case of crop resistance, genetic variation would be an invaluable resource. Similarly, if permanent forests on the moon were a goal, it would be necessary to consider carefully both which species and how much genetic variation within species would be required to make the forests resistant to either climatic change or pest outbreaks—and in the face of the inevitable uncertainty, more species would be better than fewer. A little redundancy can help stabilize neutral systems just as it can insure against failure in human-made machinery.

Thus, at a very minimum, you probably should plan to open a "branch" genetic library on the moon. On its "shelves" should be representatives of a wide variety of strains of the crop, tree, and domestic-animal species taken to the moon, as well as a selection of close relatives of the domesticated species. Once on the moon, you would need to arrange for those different strains and species to be maintained as reasonably substantial populations in relatively "natural" circumstances. For instance, it would not be wise simply to store seeds of crop plants in deep freezes. That would result in natural selection for seeds that best withstood the stress of freezing and would produce strains that were more storable but that probably had undesirable properties as well, including reduced genetic variability. Of course, if there were any chance that the lunar colony might be permanently cut off from Earth, it would be prudent to expand the lunar genetic library until it housed as many plant and animal species as possible. Then lunar scientists would have a better chance of finding new compounds for use in developing pharmaceuticals and other useful natural products.

You can see that it would be no easy exercise to determine what samples of Earth's biota should be taken to the moon. A smaller and much simpler version of this exercise, which was nonetheless instructive, was tried in the Arizona desert a few years ago. You have probably heard about the adventure called Biosphere 2 (to differentiate it from Biosphere 1, in which we are all living today). In essence, a giant greenhouse was built, covering 1.26 hectares (about 3 acres) and enclosing over 191,000 cubic meters. Within the greenhouse, which was intended to be hermetically sealed, ecosystems were constructed: grasslands, marshlands, a bit of ocean (including a coral reef), even a tropical rain forest. Plants and animals representative of all these were brought in. Soil organisms were provided by the simple expedient of importing soil from nature, and a fairly large portion of Biosphere 2 was devoted to intensive agriculture. Design and construction took about seven years and cost about $200 million. This was no fly-by-night operation.

The whole system was designed to support eight human beings. In mid-1991 four men and four women, the "Biospherians," were sealed into Biosphere 2 to stay for two years. Those of us who were cynics, who thought it was simply an experiment in human mating behavior, were wrong. Things got too tough too fast. One after another, the ecosystems collapsed and ceased providing their essential services. Unlike the ecosystems of Biosphere 1, they were unable to maintain the gaseous quality of the atmosphere. Oxygen levels dropped so precipitously that the inhabitants were living in an atmosphere equivalent to an altitude of 5,700 meters—1,000 meters higher than the top of Mount Whitney or the Matterhorn, and two and a half times the height of Australia's Mount Kosciusko. Levels of nitrous oxide spiked upward to the point where they could cause brain damage, and carbon dioxide levels fluctuated dramatically. Unbeknownst to the designers or occupants, the carbon dioxide was reacting with the concrete in Biosphere 2's structure and being converted to calcium carbonate. The oxygen needed for the reaction was being extracted from the sealed-in atmosphere. If Biosphere 2 were on the moon, decreased oxygen would have meant the doom of the entire experiment even before the two years had gone by. However, because Biosphere 2 was still enclosed in Biosphere 1, additional oxygen could be (and was) added.

Nineteen of twenty-four vertebrate species went extinct almost immediately, as did all of the pollinators, limiting the persistence of populations of plants requiring pollination to the lifetime of the colonizing individuals. Morning glories and other vines that had been added to the experiment to absorb excess carbon dioxide ran rampant. The Biospherians strove to weed them out, but their efforts fell short and the crops were overgrown. The occupants started to starve and were forced to cut down the tropical rain forest in order to plant more food crops. They also sneaked in candy bars and other rations from Biosphere 1!

Natural pest-control services were absent. As a result "crazy" ants (so named for their habit of dashing rapidly about) swarmed everywhere, and cockroaches and katydids were the other prominent survivors. Light levels in the greenhouse were lower than originally calculated, and air temperatures near the top of the greenhouse were much higher than anticipated. Since the rainfall was not adjusted to compensate, desert areas converted to chaparral. The aquatic systems accumulated nutrients that needed to be removed by growing, and harvesting by hand, large mats of algae. A planned brackish estuary had to be permanently isolated from the "ocean" because of problems managing the water chemistry. Huge amounts of electrical energy, at a cost of about $1 million each year, were required to supplement the solar energy. And there were tales of considerable friction among the Biospherians as food supplies dwindled.

In short, the Biosphere 2 experiment was an incredible disaster, but one from which many important lessons could be learned. Both our thought experiment and the Biosphere 2 breakdown underline the crucial importance of ecosystem services, and emphasize how little we know about what is required to maintain them. The services themselves are interconnected and complex, a "wild solution" to the support of life on Earth that has been produced by 3.8 billion years or more of evolution.

What basic lessons might we draw from the thought experiment and the actual Biosphere 2 experiment? (We'll skip the obvious one that self-sustaining space colonies are not going to be easy to build in the foreseeable future.) One is that we are dependent in the short term on many more kinds of organisms than it would seem at first glance. Humankind as a whole uses thousands of other species directly and is utterly dependent on some of them (like the major cereal crops) for the existence of civilization. We cannot afford to risk worldwide the kind of agricultural failure that afflicted Biosphere 2, but the chances of such failures are enhanced by factors as diverse as the genetic uniformity of major crops and the threat of rapid climatic change.

A less apparent lesson is that people benefit directly from the services that hundreds of thousands or millions of species on Earth provide. We may not use pollinators directly, but our diets would be far more restricted if they were to disappear. Decline of the birds, predatory insects, and mites, fungi, bacteria, and viruses that control crop pests and potential pests would lead to worldwide problems with maintaining agricultural production, despite the best efforts of pesticide companies. Many versions of Biosphere 2's rampant vines and crazy ants are hidden amid Earth's natural biodiversity.

Equally important, the experiments accentuate that the future of humanity is tightly intertwined with the future of nature's genetic library. That library contains the genes that could support high-yield agriculture over the long term, genes that could assist our crops in dealing with the evolution of new pests and diseases. Genes from the library could help buffer the effects of the appearance of new environmental toxins or the stresses of rapid climatic change. The library contains genetic variants and different species of trees that could help forests retain their integrity and their capacity to deliver services to humankind through otherwise catastrophic climatic events. It holds genes and species yet to be discovered as sources of crops, animal foods, pharmaceuticals, and industrial products, as well as sources of beauty, wonder, and inspiration.

The unforeseen in Biosphere 2 led to its doom as a habitat for *Homo sapiens*; the unforeseen in a rapidly changing Biosphere 1 could also contain the seeds of destruction for most of its human inhabitants—seeds that nature's library might prevent from sprouting. Life on Earth has continued for a few billion years; humanity in its current physical form goes back only a few hundred thousand years, and in its current cultural form, a mere few decades. Nature has survived without people for most of its history. Humankind clearly needs Nature and its services much more than Nature needs us.

FOUR
The Natural Internet

The Biosphere 2 story shows how connected things are in a greenhouse. If the oxygen concentration drops in one part, it drops in all parts; if ants go crazy in one part, they go crazy in all parts. Those sequences are pretty obvious. Less obvious to most people is that the entire biosphere is similarly interconnected. For example, over the period of a few months the entire atmosphere of Earth becomes completely mixed. If a cow farts in Bangladesh, the methane added to the atmosphere, by enhancing the greenhouse effect, can help change the climate in Philadelphia. John Donne once said, "No man is an island, entire of itself," but he might just as well have written, "No organism is an island, entire of itself." Every living individual is dependent on other individuals for its very existence and for the perpetuation of its kind.

That dependence is not just on parents, but on unrelated individuals that are part of the same ecosystem. The HIV virus must have a primate host in which to survive; an apple tree must have microorganisms in the soil to convert chemicals it cannot use into the nutrients it requires; a fig tree depends on small wasps that pollinate it for its reproduction; coral animals depend on tiny green algae that live within their rocky skeletons to supply them with energy; and termites, butterflies, deer, dogs, lions, chimpanzees, and people need other plants and animals to eat—and often also microbes to help them digest what they have eaten. In short, we and all other living beings are part of a natural internet but, unlike the artificial Internet we surf, our very lives depend on the natural internet's integrity.

All organisms are interconnected by vast global recycling systems known as biogeochemical cycles. If we separate the word "biogeochemical" into three parts, we see immediately what is involved: "bio" means living things, "geo" refers to the nonliving parts of the biosphere such as the soils and the atmosphere, and "chemical" indicates that we are talking about the chemistry of life, especially the fundamentals such as carbon, oxygen, nitrogen, phosphorus, and sulfur. Carbon, a basic element of life, is an especially relevant example, since it is tightly tied to the movement of that other critical substance, water (to be considered in the next chapter).

Carbon's cycle is extremely complicated, but here is an outline of what goes on (see Figure 10). The process of photosynthesis is carried out in the bodies of green plants, algae (mostly microscopic, single-celled organisms), and some microorganisms, and is fundamental to the cycling of both carbon and oxygen. Plants in effect feed themselves by taking carbon dioxide (CO_2) from the atmosphere and combining it with water in a complex process driven by solar energy. The end products are energy-rich carbohydrates such as glucose, plus new water molecules and free oxygen. When plants are eaten by animals or die, their carbohydrates (and all the molecules such as fats and proteins that make up their bodies) are broken down in a process that is the reverse of photosynthesis, a sort of slow burning called respiration. In organisms seeking to retrieve solar energy from plant (or animal) tissues, oxygen from the atmosphere does the "burning," just as it does the burning when a house goes up in flames. It combines with the carbohydrates to produce carbon dioxide and water, with a release of energy. This breakdown may occur in the digestive system of the plant eater, or that of a microorganism living on dead plant material. When an animal eats a plant, it uses the energy released from respira-

Fig. 10 The carbon cycle. Carbon dioxide in the air enters plants, and the carbon-based molecules they make are consumed by animals as diverse as cows and termites. This is the phase in which carbon is retained in living organisms. It returns to the environment by three routes: respiration (including

tion of the plant chemicals it has taken in. In the case of predators, respiration releases energy that was originally acquired from plants by herbivores. Animals use that energy to build their bodies and maintain their life processes.

Eating and being eaten form what may be thought of as the most important part of the natural internet. Eating sequences are called food chains, and they can have numerous steps. For example, a plant can fix energy from the sun by photosynthesis, and an insect can obtain that energy (and carbon, hydrogen, and oxygen) by eating the plant. A fish can then get the same energetic rewards by eating the insect, a human being by eating the fish, a lion by eating the human being, a flea by eating the lion, and a bacterium by consuming the flea. Such a food chain would have six links, excluding the links to decomposing organisms (bacteria, fungi, burying beetles, and the like) that dispose of wastes and dead bodies. The decomposers close the cycle by returning CO_2 and water to the inorganic portion of the cycle—the part that takes place in the physical environment without the participation of organisms. Eventually, no matter how many steps are in the sequence of "eat and be eaten," the oxygen returns to the atmosphere in the form of CO_2 and water, and the energy originally captured from the sun is returned to the atmosphere as heat.

But there is further complexity, even in our simplified example. Since more than one kind of organism usually dines on any given dish (hyenas, tigers, and fleas dine on people too), and each diner often devours many dishes (a fish species may consume hundreds of species of insects), food chains are normally woven into complex food webs,

---

decomposition), erosion, and combustion. Some phases can be rapid; for example, gases released from termite mounds or from the rear ends of cows may be quickly absorbed by surrounding plants. By contrast, carbon released from burning fossil fuels or the slow degradation of old snail shells on the ocean floor may take millions of years to recycle into the bodies of living organisms.

which are a vital part of the infrastructure of the natural internet (see Figure 11). While scientists tend to think of food chains and webs as fairly local affairs, they can often extend globally. For example, a species of seabird, the Arctic tern, obtains some of its energy from small fish in the Antarctic. This energy may be acquired by a scavenging polar bear when the bird returns to its breeding grounds in the northern summer. And food chain connections are only one part of the carbon-oxygen cycle. After all, that cow fart in Bangladesh contains another carbon compound, methane, and roughly one one-hundredth of the carbon in the atmosphere circulates as methane rather than as CO_2 (or, more rarely, as carbon monoxide, CO). And, of course, some of the molecules of CO_2 that you are exhaling at this very moment may well be incorporated by photosynthesis into a cecropia tree in southern Costa Rica, a malee shrub in southern Australia, or a tiny one-celled plant in the ocean off Cape Trafalgar in Spain.

Unhappily, the fact that scientists fail to understand all the linkages in the natural internet can lead to substantial difficulties. A typical case is that of red tides—plagues of toxic algae that can lead to massive fish kills and poisoning of human beings. Red tides are not new, nor are they always red. Their color depends on the pigment of the species undergoing a population explosion; they can color the ocean red, brown, or green. The occurrence of such algal blooms, mentioned in the Bible, can be inferred from the fossil record. There is evidence that the blooms are becoming more frequent, and there is no doubt that they are affecting more people because of rapidly increasing human population densities in coastal areas.

Red tides, even when not composed of toxic species, can shade submerged vegetation when they discolor the water. Thus, they alter the entire food web and cause oxygen depletion. When they are toxic, they often kill large numbers of wild fishes, many of which have commercial importance. They also can poison farmed fishes and may produce strong nerve poisons that accumulate in clams, oysters, mussels,

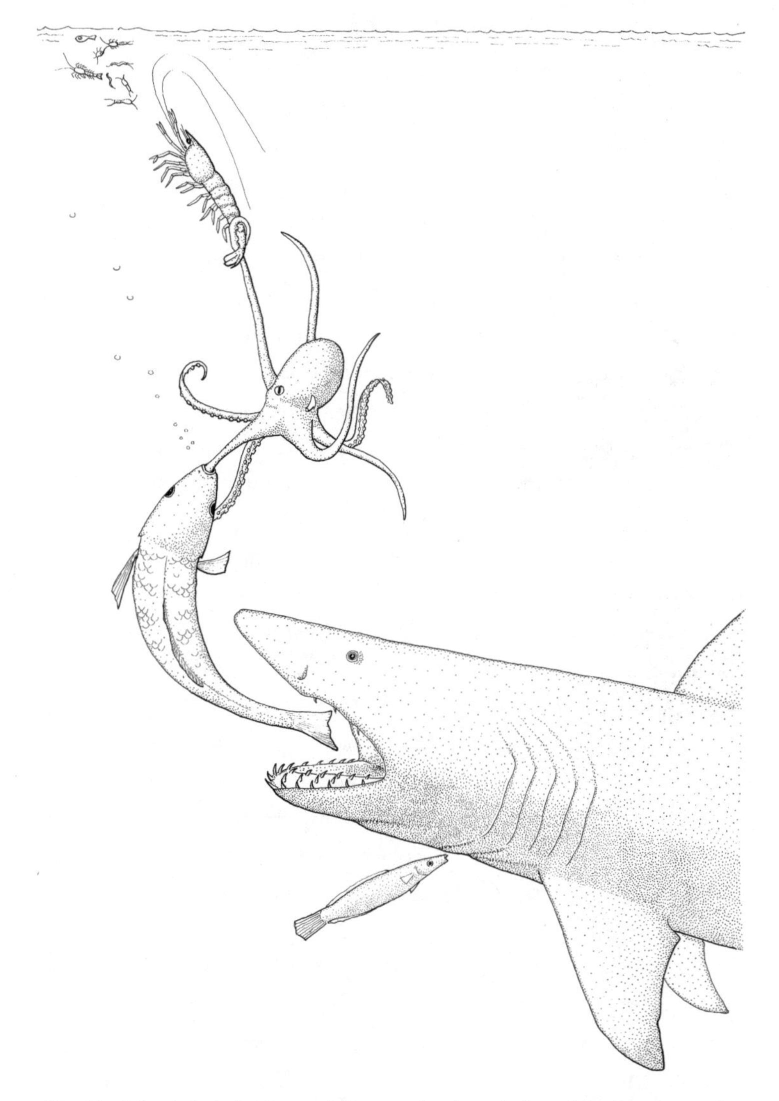

Fig. 11 A food chain in the sea. Minute animals and plants (plankton) near the surface may be food for shrimps, which may be eaten by a predator such as an octopus. Large predatory fish take octopus, and sharks, in turn, devour fish.

and scallops and make them dangerous to eat. Some of the blooms can be truly massive, affecting coastal ecosystems over many hundreds of square kilometers.

One reason for the increased frequency of blooms is certainly the flow of nutrients from farm fertilizers and disposal of human and livestock sewage through streams and rivers into coastal waters. Another appears to be the pollution of oceans by nutrients carried through the atmosphere. Nitrogen is perhaps the most thoroughly studied of those nutrients. Human activities have already doubled the amount of nitrogen entering the land-based part of the nitrogen cycle each year. About half of the nitrogen that ends up in the North Atlantic Ocean arrives from industrial plants and automobile exhausts; it also vaporizes as ammonia from ponds of hog wastes and fertilizer applications, and is released by the clearing and burning of tropical forests. It then travels on the winds to the oceans. Nitrogen pollution there triggers the explosive growth of algae, and if the species exploding is toxic, the consequences for marine life and human beings can be severe.

One instance of a catastrophic bloom involved the release of pig manure into the coastal waters off North Carolina. In June 1995 a lagoon containing more than 95 million liters of raw sewage from an intensive swine-raising operation (about twelve thousand animals) ruptured after a heavy rain. Almost all of the effluent was released, and much of it flowed several hundred meters over the land surface and emptied into a small stream about 48 kilometers upstream from an estuary. This led to an explosive growth of populations of a toxic alga belonging to a group with the sinister-sounding name, armored dinoflagellates. The species had been named *Pfiesteria piscicida* in 1991, when it was found in a North Carolina estuary associated with a kill of about a million Atlantic menhadden (a herring-like fish). The 1995 sewage spill may have killed as many as ten million fishes. The toxin released by the *Pfiesteria* caused severe psychological problems for laboratory workers exposed to it, and lesser problems for individuals who simply were on a bridge when the sewage passed underneath. Extraordinary though it may seem, this dangerous alga was utterly unknown a mere decade ago. It came to the attention of scientists because a major change in one terrestrial portion of the internet (the concentration of industrial hog farming in North Carolina) had a dramatic impact on a marine part of the internet (the estuaries of the state).

Another example of how difficult it is to understand the intricacies of the natural internet comes from beaver ponds. When a dam is abandoned by its makers, it begins to fill in with silt, at some point its walls crumble, and the water drains away. The mud left behind is quickly colonized by grasses, sedges, and other small plants, and in many cases tree seedlings soon follow. But sometimes they do not, and scientists have puzzled over why some abandoned beaver ponds take as long as a century to revert to forest. An answer may be emerging, at least in forests where the dominant tree is the black spruce. This tree depends on a highly specialized fungus that wraps around its rootlets and helps absorb nutrients. Decades of submersion under the beaver pond may gradually drown the fungus, so that when the water drains away none is left in the mud for the spruce saplings. Experimental tree seedlings planted in old dams without these fungi did not thrive. So it seems that a minute fungus is one node of the forest internet.

Enter another, in the form of the red-backed vole that forages around the edges of forest clearings such as old beaver ponds. This small mammal has a varied diet, and its feces contain the spores of the same fungus. Experiments showed that when tree seedlings are planted in old dams, together with vole droppings, they do just fine. Anyone walking through a spruce forest may be completely unaware of either the fungus, which lurks underground, or the vole that scuttles out of sight, but both are vital parts of the internet of species that form the forest.

While on the subject of forest trees and the special fungi on which they depend, let us explore the nature of this relationship. In this case, the internet is physical; that is, the roots and fungi are actually connected. The fungi link the roots of the same and different species of trees below ground; they are also pipelines for the exchange of nutrients. You may have wondered how small trees shaded by larger ones ever receive enough sunlight to make food via photosynthesis. The answer is that sometimes they do not; the nutritious carbohydrates are transferred to the smaller trees from the larger ones that have leafy canopies up in the sun. In addition, sometimes small parasitic plants sit on the forest floor with their roots firmly embedded in tree roots, being fed by the photosynthesis carried out high above. Many years ago an old forester said that trees talk to one another. Some thought he was daft, but research on the root-fungus internet has shown that he was right.

From the standpoint of what other species do for us, the "eat and be eaten" structure of the natural internet is very important. A great many of the useful items we extract from Nature are either chemicals that organisms have evolved to defend themselves against being eaten (quinine is an example), or that organisms use to obtain food (spiders' silk). In this process, called coevolution, the evolution of one species prompts an evolutionary response in another. In this case, one kind of organism is evolving ways to avoid being eaten by another, and the other is attempting to evolve ways around any defenses the first manages to acquire. It is our belief that the evolution of plant defenses is one of the most critical parts of Darwin's panorama. Plants have a serious problem in that they cannot run away from the animals that want to eat them. "Stand and fight" is their motto, a slogan familiar to anyone who has had a close encounter with a cactus, poison ivy or poison oak, or a green apple. While many of our green friends have adopted mechanical defenses such as spines or tough bark, many more have opted for chemical warfare. As we shall see in later chapters, human beings have adopted plant chemical-warfare weapons to meet a variety of their own needs.

Coevolutionary relationships can make organisms especially vulnerable to human-caused disturbances. For example, many coral reefs today are dying because of "coral bleaching." While coral reefs are hard and feel like rock, they are secreted by tiny polyps, inside which reside algae that, through their photosynthetic activities, supply crucial nutrients to the coral animals. This coevolved relationship is intimate enough to be called a symbiosis. The algae, however, are very sensitive to increases in temperature. The gradual warming of the oceans, presumably connected to a global warming, is killing them. Without their algae the corals die, and eventually the reefs disintegrate—destroying the economically valuable biodiversity associated with them, including the rich fisheries and their protective function for shorelines.

Many organisms aid us simply by the way they feed or are fed on. For example, in the process of eating their fill, decomposers not only provide the nutrients needed to keep our food coming, but also dispose of all sorts of obnoxious wastes that would sooner or later make our lives miserable. Microorganisms make critical contributions to our nutrition, not only by supplying nutrients to the plants that we (and the animals we eat) feed on, but also by living in our intestines and helping with human digestive processes, often by providing nutrients our own systems cannot extract from our food.

Of course, not all organisms are beneficial to us. We are connected to the global internet by being on the menu of many creatures besides fleas and lions. We provide nourishment not only for occasional big predators and a lot of burrowing and blood-sucking insects, ticks, and mites, but we are also hosts to myriad hungry microorganisms that cause disease. Just as the plants are coevolving with *their* attackers, so we are coevolving with *ours.* Medicine in essence attempts to sever some of those connections to the internet, and we often turn to products derived from biodiversity to help us counter the assaults of other elements of biodiversity.

For instance, the variola virus that caused smallpox was eventually conquered through use of the vaccinia cowpox virus, which caused a related disease in cows known as cowpox. Smallpox was a horrendous disease that afflicted humanity for at least three millennia, perhaps originating as a mutation of an animal virus such as cowpox. In Europe in the seventeenth century, more people died annually of smallpox than of leprosy, the black plague, or syphilis. As late as 1950, smallpox was killing more than a million people a year in India, and it was still a threat in the late 1960s to almost two-thirds of Earth's population. It killed one in five victims, often scarring or blinding those who survived. A determined global vaccination program has since made the smallpox virus extinct in the wild. Since there was no other animal in which the smallpox virus could survive, removing us as a node of the internet (through vaccination) actually made that virus an endangered species. There is now a substantial debate over whether to preserve laboratory examples of the virus or to destroy the cultures. The basic issue is whether the risks of the virus escaping (or, more likely, being deliberately released as a weapon of biological warfare or terror) and once again infecting the human population outweigh the benefits of retaining it as a research tool. And, of course, there is always the chance that the smallpox virus will be recreated by the same mutation or series of mutations that produced it in the first place.

Feeding relationships also influence the overall distribution patterns of the plants and animals with which *Homo sapiens* interacts. Some of these relationships are not obviously beneficial. For instance, cattle cannot be raised in large areas of Africa because those areas are also home to tsetse flies that love to feed on our bovine companions. While individual flies do not eat much, they do transmit a deadly protozoan parasite (a trypanosome), related to the ones that cause sleeping sickness in human beings, that give the cattle a fatal disease called nagana. But the trypanosome is not the end of the story. Cattle are grazers; they eat grass and were originally animals of the moist tropics.

Where human beings have introduced them, the herds have often contributed to the creation of deserts in areas that once were savannas. There are at least a couple of reasons. First, cattle require daily rations of water and thus must parade back and forth to water holes, trampling the vegetation and compacting large areas of soil. Second, the moist droppings of cattle rapidly lose ammonia to the atmosphere. As they dry, they heat up in the sun and kill the grass beneath them, forming a nearly impenetrable soil covering (sometimes called a fecal pavement) through which grasses have difficulty sprouting. Cattle also prefer certain grasses and graze them heavily. Thus, not only is an area stocked with cattle threatened with desertification, but the composition of the flora shifts toward grasses that cattle find less desirable.

In contrast, antelope, giraffes, zebras, warthogs, and other African herbivores can utilize a wide variety of savanna plants. The native African herbivores of an area tend to graze and browse (eat leaves, twigs, and young shoots from shrubs and trees) a variety of plants and do not cause sharp shifts in the flora of the range. Native herbivores such as antelopes, in contrast to cattle, produce dry fecal pellets that are readily broken down by decomposers, returning their nutrients to the soil. The point is that the presence of one tiny trypanosomal parasite, one node of the natural internet, generally excludes cattle and therefore tends to maintain a rich ecosystem that contains a wide diversity of wildlife. Where the parasite lives, savanna stays savanna and does not become desert.

The global internet is kept in place by a plethora of mechanisms. Therefore, if an area of natural habitat is damaged or seriously degraded by, say, a landslide, tidal wave, or forest fire, the web of life in the area will be reestablished. A major mechanism in this case is transport of organisms from one location to another. While the wind carries the seeds of some plants from place to place, as it does some animals such as small birds, various insects, and "ballooning" spiderlings (baby spiders that have spun long silk threads into the wind, which then carries them aloft), many other organisms may be involved in the transport process. Waterbirds often carry aquatic plants and microorganisms from pond to pond on their bodies. Birds and bats are famous for dispersing seeds of plants in their droppings. Indeed, when attempts are made to reestablish tropical forests in previously logged areas, often poles are erected with crossbars to serve as bird perches. Birds then fly out from the forest, alight on the poles, and defecate, seeding the pasture around the pole.

Shrikes make one of the strangest connections yet found in the natural internet. These small, carnivorous songbirds help disperse the seeds of a plant in the tomato family on one of the Canary Islands. The shrikes do not eat the fruits of the plants, but lizards do. The shrikes in turn eat the lizards, and thus acquire the seeds that are then passed out of the shrikes in their feces. The digestive systems of the bird and the reptile may be very different, or the seeds may linger in the digestive system for less time when a shrike eats a seed-containing lizard than if the lizard is allowed to pass the seed in its own good time. Although we do not really know, the fact is that seeds from shrike droppings are more likely to germinate than those from either lizard droppings or uneaten fruit. A carnivore might disperse seeds over barriers (such as the ocean between islands) impassable to the herbivore that originally eats the fruits. If we did not have such diverse and numerous transport systems, the recovery of damaged sections of the internet that supply society with critical services would be much delayed—and in some circumstances would not occur at all.

There is more to the connections in the natural internet than simply the links forged by feeding—by herbivory, predation, and decomposition. Many animals, plants, and microorganisms are linked by competition for the same resources. Obvious examples are our battles with the insects, mites, fungi, weeds, and other pests that compete for our crops. The entire internet is in some respects shaped by competition. In many cases, two closely related organisms are unable to persist in the same area because one is a superior competitor and drives populations of the other to extinction. This process has been documented for many pairs of bird species on islands in the tropical Pacific and for *Anolis* lizards in the Caribbean, among others. Or one species may be confined to a limited habitat because of the presence of another, as when green sunfishes are concentrated in vegetated areas because bluegills and largemouth bass exclude them from a lake's open waters.

More speculatively, competition from diverse and dominant dinosaurs may have held primitive mammals to a relatively obscure portion of Earth's fauna during the Cretaceous period. Then an extraterrestrial object appears to have struck our planet some sixty-five million years ago, wiping out the dinosaurs and ending the Cretaceous. With their main competitors gone, the mammals underwent enormous proliferation (technically known as an "adaptive radiation") and filled the niches left by the dinosaurs. Were it not for that chance event, the natural internet would look very different today; a group of *Velociraptor* (those smart dinosaurs from Jurassic Park) descendants might be writing and reading this book!

Another event that has dramatically changed the natural internet is the appearance of *Homo sapiens.* The rise of our species to global dominance is the single most remarkable terrestrial event in the sixty-five million years since we said goodbye to the dinosaurs. Human beings as a species are only rivaled in total weight by other species, such as cattle, that they have domesticated. There are now six billion of us, some five times as many as there were a hundred fifty years ago; and because every one of us longs for a higher standard of living, there is a huge increase in the environmental impact of each person. The total human assault on the natural internet has increased more than twenty-fold in the same period. As a result, humankind is now significantly changing the gaseous quality of the atmosphere, mobilizing various minerals at many times their natural rates, vastly altering the land surface of Earth, decimating oceanic fish stocks, and polluting the entire planet with toxic substances.

Perhaps the biggest single threat to the integrity of the natural internet today is rapid climatic alteration. If *Homo sapiens* is unlucky, the changes it is making by adding greenhouse gases to the atmosphere and altering the reflectivity of Earth's land surface could lead to unprecedented weather patterns. A natural internet will continue to exist, but it will be very different—one that may have fewer interacting parts because of high rates of extinction, and one that may have new properties to which human populations would have difficulty adjusting. Imagine how the Biosphere 2 experiment would have evolved if, in addition to its other problems, the whole unit had to face unprecedented changes in temperature, airflow, and crop-watering regimes.

Overfishing is a significant threat to the natural internet. It does more than simply harm fish stocks; it also makes serious alterations to the oceanic part of the internet. The competitive and predatory relationships of the net are changed, frequently in ways that are not entirely understood. For instance, excessive harvesting of the California sardine caused the stock to crash; it has never recovered, presumably because of changes in the internet. Even more serious are the effects of trawling. Heavy nets are dragged over the ocean floor, in many places more than once a year. The effect is similar to that of clear-cutting a forest but, hidden beneath the ocean waves, the damage does not attract the same attention or publicity. Trawling destroys sponges, corals, bryozoans, and other bottom organisms that feed on materials suspended in the water, and could alter the internet by reducing the productivity of fisheries.

Homogenization is another dramatic effect of human activities on the internet. Human beings have taken so many species from one part of the world to another that we can encounter many of them no matter where we are. And many species, when taken to an entirely new part of the world, turn into weeds or ferals. They become plants or animals that have no natural enemies and so breed in vast numbers, often on the scale of a plague. By transporting organisms around the globe we cause biotic invasions and have created a series of disasters on local and regional scales. We introduce predators to habitats where none existed previously. The native animals of many islands have been exterminated by rats, cats, goats, and pigs toted around the world on ships. We also carry plants beyond the reach of their natural enemies. A huge portion of Australia was made unavailable to farming and grazing after a South American cactus was imported as an ornamental plant. Hardy African *Imperata* grass reduces the usefulness of pastures in Costa Rica and elsewhere in the tropics, and in the intermountain west of the United States, inedible imported *Bromus tectorum* (cheat grass) plays a similar destructive role. We transport tough competitors and dangerous diseases to areas where native species have had no evolutionary experience of them. Hole-nesting birds in North America, for instance, have their populations threatened by aggressive starlings imported from Europe. And mosquitoes that traveled around the Pacific with Polynesian colonists carried diseases that helped to exterminate most low-elevation native Hawaiian bird populations.

Of course, the travels of human beings around the world long before the dawn of history also had a significant impact on the natural internet. Prehistoric human use of fire altered landscapes over much of the globe, and appears to have created savannas in areas of Africa that previously were woodlands. The invasion of the Western Hemisphere by *Homo sapiens* thousands of years ago led to the demise of the so-called Pleistocene megafauna—numerous species of large mammals such as giant beavers, giant capybaras, flat-headed peccaries, Shasta ground sloths, long-horned bison, saber-toothed "tigers," and wooly mammoths. The demise of those animals in turn altered food webs and led to important changes in the flora of North America.

Current human activities are rapidly changing the natural internet of many areas; for instance, in the eastern forests of North America suppression of natural forest fires is leading to substantial changes in plant communities. Before forest-fire prevention activities related to the "Smoky the Bear" programs of the U.S. Forest Service, red maples were largely confined to swamps and other wetlands. They were killed in drier areas by periodic natural (lightning-caused) forest fires. Those fires were much less likely to kill hickories and oaks, which are fire resistant because of their thick barks and deep roots. In addition, neither deer nor gypsy moths, major predators on oaks, feed readily on maples. The maples are shade tolerant, seed earlier than the oaks or hickories, and thrive in disturbed areas and on a wide range of soil types, giving them an additional advantage over oaks and hickories. As a result, the maples are taking over what were previously known as oak-hickory forests. The negative effect is felt by deer, squirrels, jays, and various insect species that feed on oak or hickory. The loss of oaks is particularly important because of the dependence of many animals on acorns, especially since an imported disease has largely exterminated chestnuts, which supplied a similar food. So by "preventing forest fires," we actually do much more. In fact, we are substantially altering a section of the natural internet, an alteration that is hardly to our advantage. For the soft maple wood is not as valuable as the oak and hickory hardwoods, thus impacting the timber industry. The further consequences of fire suppression are unknown at the moment, but unless a system of controlled burning is instituted, we are virtually certain to find out. In the process, the character of northeastern forests may be altered for millennia.

The natural internet supplies us, through delivery of ecosystem services, with "wild solutions" that help to answer some of humanity's most urgent problems: how to obtain nutritious food, how to maintain a reasonable range of temperatures in which to live, how to have clean air to breathe and freshwater to drink, and how to avoid severe weather. We unravel the natural internet at our peril.

FIVE

Thirst: Ecosystems and the World's Water Supply

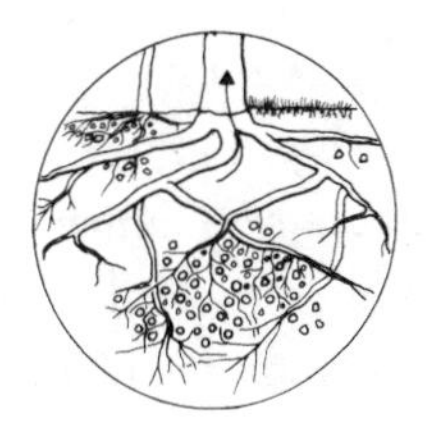

"Water is the best of all things," said Pindar, the Greek lyric poet some twenty-five hundred years ago. How right he was! We cannot do without it, and few pleasures are more satisfying than a cool drink from a mountain spring when we are truly thirsty. But perhaps Pindar should have said that *fresh*water is the best of all things, since only about 2.5 percent of Earth's water is fresh. The rest is salt water, which will hardly slake one's thirst, as the Ancient Mariner noted. Moreover, of that small fraction of our water that is fresh, some two-thirds is totally inaccessible in ice caps and glaciers. Only about 0.77 percent is in lakes, swamps, rivers, aquifers, pores in soils, the atmosphere, and the bodies of living organisms. But Earth has correctly been called the water planet, and as a result that 0.77 percent is a huge amount of water—more than 10 million cubic kilometers.

The renewable freshwater supply, that flowing through the solar-powered hydrologic cycle, is much less. The cycle lifts about 430,000 cubic kilometers of water by evaporation from the surface of the oceans. Roughly 390,000 cubic kilometers of that water fall right back into the oceans as precipitation. But some 40,000 cubic kilometers are carried by winds from the oceans to the land surface, where they add to the rain, snow, sleet, and hail that fall there. That water with its source in oceanic evaporation, accounts for a little over a third of terrestrial precipitation; a little less than two-thirds has as its source water evaporating from the land surface and pumped out of the ground by plants.

Those sources together account for the roughly 110,000 cubic kilometers of water that fall as precipitation on the land surface annually. That is the most important to humanity, even though it is only about 1 percent of the freshwater stored in groundwater, lakes, rivers, and elsewhere (see Figure 12). Of the renewable 110,000 cubic kilometers, only a third or so runs back to the sea in rivers, streams, and groundwater flows. And only about 10 percent—some 12,500 cubic kilometers—is reasonably accessible runoff. That much sloshing liquid may seem like a vast supply, except that humankind is already appropriating more than half for its own use. Furthermore, the human population *and* its per capita demand for freshwater are both still growing.

Two of the key roles wild organisms play in the hydrologic cycle (from the point of view of humanity) are helping to maintain quality and supply. Plants and microorganisms play a major part in this process, as they do in other cycles of the natural internet. Vegetation slows the flow of water over the land surface toward stream systems and thus provides time for it to soak into the ground. Plants break the force of heavy rains and protect the soil. This is particularly obvious in tropical rain forests where torrential downpours are intercepted by the tree canopy so that, although the ground is very wet, the water soaks gently into the leaf litter and soil. In addition, the roots of plants in a well-vegetated watershed both hold the soil in place and help to retain nutrients in the system. Soil secured in this way absorbs rainfall and then gradually releases the water into streams and rivers. The result is an even flow that is generally beneficial to humanity—making it easier to use water for transport, power generation, irrigation, recreation, and the like. When soil is eroded away, this water-metering function is

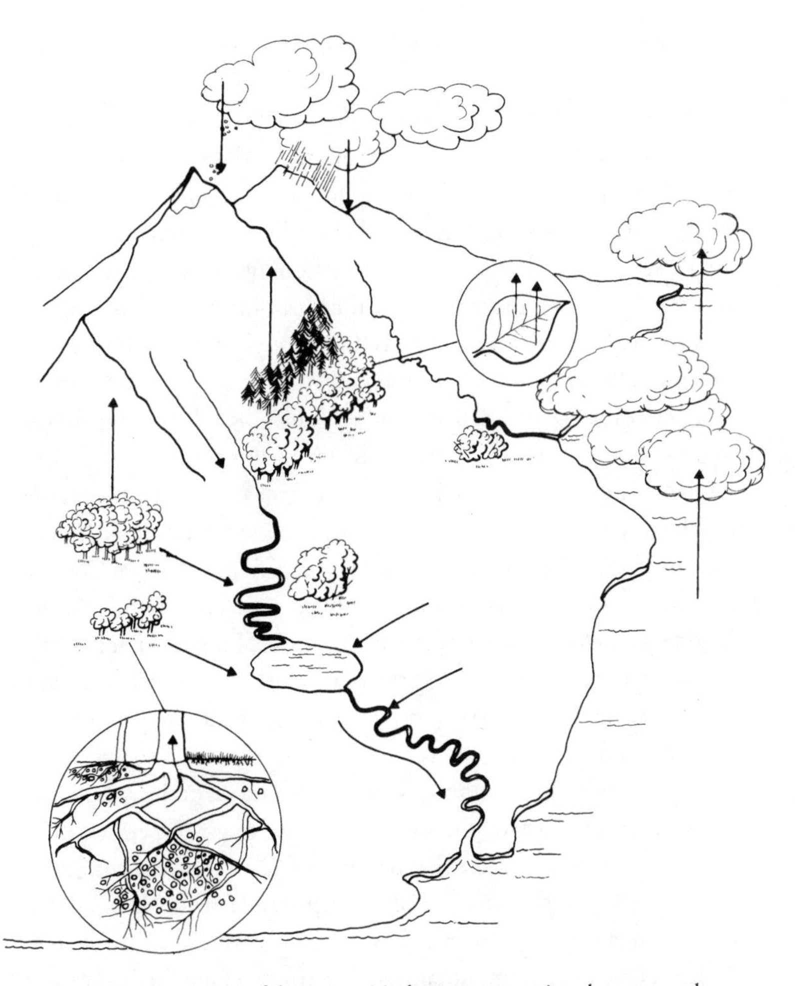

Fig. 12 Components of the most critical ecosystem service, the water cycle. Water evaporates from the oceans and the land to the atmosphere and returns through precipitation. Some water may linger in underground aquifers and polar ice caps for very long periods.

impaired and the results downstream can be alternating droughts and floods.

The loss of the flood-control service supplied by plants in natural ecosystems is demonstrated dramatically in southern California every year or so. Much of the vegetation in the coastal region is chaparral. It consists of many species of shrubs that are highly flammable so that, after a season of dry weather, it tends to burn fiercely over large areas. In the winter, when the rain comes, the burnt-out areas are inevitably plagued with mudslides and floods, for the vegetation has not yet recovered sufficiently to control the flow of water. In the midwestern and southeastern portions of the United States, disastrous spring floods are frequent. One reason is vast upstream areas where agriculture has removed much of the native vegetation that once broke the fall of rain and held soils in place. It appears that the most famous flood/mudslide disaster of the late 1990s, caused by Hurricane Mitch in Central America, was made worse by the removal of natural plant growth from hillsides needed for cultivation. Similarly, areas of India and Bangladesh are more subject to inundation than previously, owing in part to deforestation in the Himalayas and the resulting diminution of the flood-control ecosystem service.

A striking example of the way in which wild plants serve humanity by assuring water supplies is provided by a tragic "experiment" in Rwanda, carried out before that nation suffered its horrendous genocidal strife late in the twentieth century. Rwanda's Parc National des Volcans includes most of the famed Virunga volcanoes of central Africa. These volcanic mountains shelter groups of mountain gorillas, some of the last on the planet. In the 1970s the park, which made up less than 0.5 percent of Rwanda's land area, supplied about 10 percent of that overpopulated nation's agricultural water. It acted as a gigantic sponge, imbibing rainfall and slowly feeding it into streams for the use of thirsty farm fields and people. But the volcano park was yet more important prior to 1969, when it was two-thirds again as large. That year a substantial portion of the park was cleared to grow the daisylike flowers from which pyrethrum, a natural insecticide, is extracted. By the time the pyrethrum scheme proved to be a failure, the forest was already gone, and with population pressures on the land increasing there was no opportunity to try to restore it. Some streams dried up as a result and, to make the economic failure doubly disastrous, the habitat for the gorillas was reduced and their survival was threatened. Rich tourists who want the incredible experience of an intimate visit with wild gorillas have been an important source of foreign exchange for Rwanda.

Microorganisms, especially bacteria, help to purify water by breaking down wastes, playing the role of decomposers in the natural internet. Where there are few people, streams and rivers can be kept relatively pure by natural action; but if the waste load becomes too high, the decomposers cannot handle the task and the water becomes unfit to drink. Unfortunately, in much of the world the density of human beings is such as to overwhelm the natural purification capacity of lakes and streams, even for contaminants that can be disposed of by microorganisms. Furthermore, there is widespread pollution from synthetic chemicals decomposers either cannot break down or only decompose with difficulty. Basically, decomposers fail to recognize the pollutants as food. Those chemicals, and dangerous pathogens like the single-celled *Giardia,* make almost all untreated water suspect. In the high mountains of Colorado surrounding the Rocky Mountain Biological Laboratory, where the three of us did research in the 1970s, scientists were able to drink the water in streams with impunity. Now the danger of *Giardia* is too great; the scale of human intrusion has overwhelmed the natural dilution and purification once provided by stream ecosystems.

Of course, humanity does not just use the portion of the annual precipitation that becomes runoff. It also uses part of the water that is returned to the atmosphere through evaporation and transpiration by plants. In transpiration, plants pump water out of the ground and return it to the atmosphere through tiny pores in their leaves. These pores also provide entry points for carbon dioxide, which is essential to the process of photosynthesis. The pores lead into the moist interior of the leaves and, as a result, water inevitably escapes through them. The leaves of plants have very little structural strength and remain turgid because of the water pressure in their cells. Thus, while the pores are open it is essential that the plants continuously extract water from the ground. If the plants run out of water or die, transpiration ceases and the leaves droop; the plant "wilts," as gardeners say. Transpiration is also the process that gives nonwoody plants their stiff structure: a dead tree will stand upright for a long time, but a dead petunia will collapse into a miserable heap as its cells dry out and lose their turgor.

A large amount of water can be involved in transpiration. One Douglas fir can transpire 100 liters of water in a single summer day. An average corn plant has about 250 liters of water pass through it in its lifetime, and it takes about 800 liters of water to produce a single pound of dry rice. Combined, evaporation and transpiration remove about 70,000 cubic kilometers from the land surface of Earth each year. In a single hour after a downpour, about 10 tons of water can rise from one hectare of an evergreen forest. That rate cannot be maintained, but about 50 tons of water can leave that same area by evapotranspiration in twenty-four hours. Some trees and deep-rooted shrubs play an especially important role in making water far down in the soil available to the hydrologic cycle. Indeed, plants capable of pumping such water often leak some out of shallow rootlets into dry soil. There the water serves as a ready source for the pumping plant, but it also can be used by more shallow-rooted plants.

The water that passes through plants is not only critical to them, it is also critical to us. It brings with it the nutrients that permit plants to develop, and that are essential to our own growth and maintenance.

All the nutrients we need are supplied directly by plants, or are passed on indirectly by moving up the food chain through plant-eating animals to the other animals (herbivores or carnivores) that we consume. Most of our animal foods come from herbivores, but many of the fishes we eat have positions higher on the food chain. Evapotranspiration from human-dominated land, employed in uses ranging from growing wheat and grazing cattle to harvesting wood from tree farms and cultivating lawns, is already estimated to be about a quarter of the total. And, as in the case of runoff, the evapotranspiration portion of the hydrologic cycle will be under pressure from continuing expansion of the human enterprise.

The importance of natural systems in supplying water to humanity is underscored by the history of the New York City water supply. In the nineteenth century the city's water was so pure, and had such a fine reputation, that it was bottled and sold throughout the northeastern United States. Then population growth, inadequate sanitary facilities, and heavy use of agricultural chemicals changed the city's watershed in the Catskill Mountains. Gradually, enough pesticides, fertilizers, and sewage entered the water supply that its quality fell below the standards set by the U.S. Environmental Protection Agency. Human activities overwhelmed the water purification service supplied by soil microorganisms, plant roots, and the filtering and sedimentation functions of natural soils and swamps.

Prior to this damage, water from the forests of the watershed was provided free. Afterward, however, the city had to confront the cost of replacing this service. Construction of a water treatment plant was estimated at $6 billion to $8 billion, and the annual operating costs were projected to be in the vicinity of $300 million. That put a minimum value on the ecosystem services supplied by the natural community of the watershed. Other services, such as carbon sequestration to aid climate stability and provision of recreational features, were not even considered. Faced with the enormous costs of the treatment plant, New York City made a supremely sensible decision: it decided to invest in natural capital rather than man-made capital. Instead of constructing the plant, the city chose to restore the natural purification mechanism and prevent it from being overloaded. It could do so by purchasing land that could be left in a natural state or have restricted usage, by subsidizing replacement of faulty septic systems, and so on. The cost of reducing pressures on the ecosystem to the point where it could once again deliver high-quality water was estimated to be at most $1.5 billion. Refurbishing natural capital required only a quarter of the funds that would need to be invested in human-made capital to achieve the same result. And, of course, many other economic benefits would accrue from that plan of action.

Having made the decision, in 1997 New York floated an issue of "environmental bonds." The proceeds were to be spent on restoration of the Catskill ecosystem's water purification function. It was, in every way, a "wild solution" to a serious environmental problem—and with luck it will again make New York City's water supply "the best of all things."

SIX

Garbage Red in Tooth and Claw

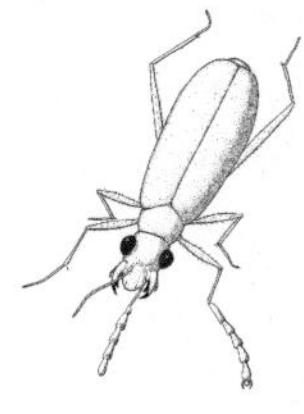

All the time, everywhere, all the species in the world are turning resources into waste, and that waste has to go somewhere. In fact, as we all know, it invariably ends up in the environment: in the soil, in a river or lake, in the sea, or in the atmosphere. The truth is that waste materials—whether they come from an ant colony, a clump of trees, a coral reef, a refrigerator factory, or our kitchen—end up in the back yard of some other organism. For hundreds of millions of years, natural wastes have been dumped on the land and in the sea and, for the same period, organisms that use those wastes for food have evolved and proliferated. Those organisms are immensely successful and come in a bewildering array of shapes, sizes, and abilities. By abilities we mean the different ways of converting wastes into resources that other creatures can use. Consider the wastes of a forest—dead leaves and flowers, logs, branches, twigs, and roots—generated all year long for thousands of centuries. Without the waste-eaters, technically known as decomposers, all that trash would have piled up and suffocated the forest long ago.

The waste from modern human societies is a vast addition to the natural load. Then too, it contains a variety of highly toxic substances and materials strongly resistant to degradation. One estimate of the organic wastes produced annually in the United States from human sewage, livestock manure, urban and industrial materials, plus wastes from agriculture, forestry, and food processing is 6 billion metric tons.

A dramatic illustration of the sheer quantity of human wastes is the massive garbage dumps of New York City that have accumulated over two centuries. These dumps are large enough to be a conspicuous feature of the east coast of North America. Decomposers in the form of bacteria are attacking the waste, but it is going to take a very long time to get through all these materials. Furthermore, as many are the products of very recent sophisticated manufacturing processes, they are new to the planet and thus create a sudden and daunting challenge to waste-eaters that have never encountered them before. Most of us think seldom, if at all, about decomposers, yet their activities keep wastes, both natural and manmade, from crushing us all. In this chapter we examine waste-eaters and show how they will help clean up the modern environment and even give rise to new industries.

All food chains end up the same way because, sooner or later, the natural decomposers take over. Imagine a food chain with the following links: a grass grows, then flowers and sets seed. Some seed is eaten by a mouse, which in turn is eaten by a snake. Then a hawk, out hunting, captures and consumes the snake, bones and all. Now visualize another kind of food chain in the ocean: microscopic algae, each consisting of a single cell, float on the current only to be eaten by a tiny shrimp as it filters the water for food particles. The shrimp is eaten by a small fish, which is later swallowed by a larger fish, which in turn is hunted down by a seal. Finally, the seal becomes the prey of a killer whale. So far, there has been no mention of decomposers. But, eventually the hawk and the killer whale will die, possibly of disease or certainly of old age. Enter the decomposers, an army of them: bacteria, fungi, and a variety of small animals that use the corpses as food. They keep some of the materials to build their own bodies and release the rest into the environment, where they are ultimately recycled as food by new generations of organisms.

Before we move on, let us backtrack to the decomposers, which have in fact been involved all along. Not in the obvious role of getting rid of the corpses, but in the mundane business of breaking down the dung and other wastes of all the species involved in the food chain. To return to the grass, some of its leaves and seeds may be damaged by insects or snails and fall to the ground into the waiting arms of the decomposers in the soil. Several times a day the mouse grooms itself; its hair, saliva, and fragments of discarded skin become meals for the decomposers. So does its dung, to which we shall return in a moment. The cast-off skin of the snake is decomposer food; if the snake is unlucky and becomes sick, perhaps from an infestation of ticks, it will die and its body will decompose. And the hawk produces several kinds of nutritious decomposer food, such as old feathers, regurgitated fur and bones from the prey, dung of course, and in many cases the corpses of young nestlings unable to compete with their older nestmates to get a fair share of the food brought by the parents.

It will be obvious by now that many items we might regard as waste in fact constitute food for decomposers. The amount of hair falling from a grooming mouse or the quantity of food in the husk of a discarded seed may seem small, but all the hair and husks from all the mice and seeds in, say, an entire valley quickly add up to a lot of decomposer resources. When we consider dung, the quantities can be truly awesome. The amount of excrement produced by all organisms in the world, every day, is vast. Yet other than on the farm it is a neglected subject. Indeed, one way of appreciating decomposers is to visit a farm, particularly toward the end of winter when months of cow dung have accumulated, knee deep, in the feedlots. The smell indicates that the decomposers are already at work, and that when spring arrives the dung will be spread across the fields as fertilizer. Now just imagine a total absence of decomposers with another winter coming on and the level of dung already rising. . .

Fortunately, dung is the staff of life for hundreds of thousands of decomposer species, bacteria, fungi, single-celled animals, and all kinds of invertebrate animals such as roundworms, earthworms, and

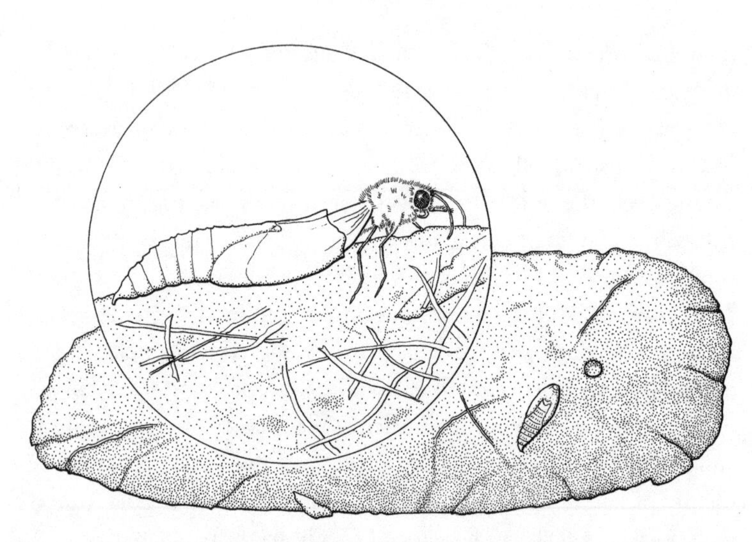

Fig. 13 A lump of koala dung is home to this moth, a highly specialized decomposer. Under magnification, an adult is seen hatching from a pupa. To the right, an unmagnified pupa is visible, together with the hole through which the larva emerged so as to pupate on the surface.

the larvae of countless species of flies and beetles (Figure 13). There are even bees that harvest dung rather than pollen. Koala dung in Australia and sloth dung in Central America each have their own, highly specialized dung-eating moth species, and throughout Africa hundreds of kinds of scarab beetles roll dung balls for their larvae to eat. Eyewitnesses in East Africa have observed dung beetles homing in on wild buffalo dung as it hits the ground. When you consider that a full-grown elephant produces several pounds of dung every day, we are fortunate that each heap is eagerly awaited by a variety of scavengers that break it down within forty-eight hours. Dung disposal is a quick and efficient service in most parts of the world, as we will emphasize later in the chapter.

Sometimes waste is obvious; think of a beached whale or a dead bird on the lawn. At other times it is far harder to appreciate. For example, if you lie quietly under a tree in hot weather, you often see a very light rain of tiny particles, illuminated in the shafts of sunlight. Many of those particles are insect dung and each one, although infinitesimal, is a full meal for some bacterium or fungus. We are also the providers of a light rain of food in our own homes, in the form of minute fragments of skin and hair that fall from our bodies. The process is completely natural and healthy, but often these bits are consumed by tiny mites that cohabit with us, largely unseen, as domestic decomposers. One unfortunate result of this food chain is that the dung of the mites, when airborne, can cause asthma; part of the cure, then, is to eliminate the mites.

Decomposers convert materials that have been used to materials that can be used again—in other words, they are recyclers. Transforming the waste of the world is a task accomplished by a vast number of individual organisms that belong to a huge variety of decomposer species. To understand this world that cannot be seen with the naked eye, we will look at the decomposers that live in soils. They are probably the best understood, in that their activities are crucial to agriculture, horticulture, forestry, and gardening. For purposes of study, we divide them into four groups: the microbes (bacteria and fungi), the microinvertebrates, the mesoinvertebrates, and the macroinvertebrates.

Soil Microbes

Before examining soil bacteria, we might find it useful to have a brief look at some global figures. A group of scientists in the United States estimates that there are roughly 5×10^{30} or 5,000,000,000,000,000,000,000,000,000,000 bacterial cells in the world at any given time. The weight of carbon they contain is probably close to the weight of the carbon in all plants throughout the world, forests included. (Carbon is an essential part of every organism, for it combines with hydrogen and oxygen to form the organic molecules that build, maintain, and control cells.) The total amount of nitrogen and phosphorus in the bacteria (remember these are precious chemicals, required as fertilizer, that must be recycled) greatly exceeds the amount in all plants. To drive home the message that bacteria are everywhere in unimaginable numbers, the same scientists estimate that there are 3×10^8 or 300,000,000 bacterial cells on the skin of the average person. Add those in our intestines, and the bacterial cells far outnumber the human cells in our own bodies!

The soil is a prime habitat for bacteria, and soil bacteria come in many different shapes, including spheres, rods, and corkscrews. Some can move under their own power; many move only when disturbed by water or other passing organisms. One gram of topsoil (enough to cover your fingernail) normally contains between a hundred million and a thousand million bacterial cells, depending on its acidity and organic content. Fungi are much harder to count, as their bodies are not usually separate cells like bacteria, but are long, microscopic hair-like filaments that penetrate the soil, absorbing nutrients as they grow. A fungal filament has much the same purpose as a column of foraging ants: to explore every nook and cranny of the microlandscape for food. It is possible to estimate the length of fungal filaments at several hundred meters for each gram of soil. Fungi are generally classified as microorganisms because their filaments are microscopic. Sometimes, though, their reproductive structures (such as those of mushrooms, toadstools, and shelf fungi) are large, colorful, and easily seen. However, there is another aspect to the matter of size in the fungi, because vast numbers of microscopic filaments can add up to very large organisms, independent of reproductive structures. For example, the filaments of one fungus growing in the forests of Michigan cover a subterranean area of about 12 hectares, are estimated to weigh about 100 tons, and apparently are fifteen hundred years old. It is hard to imagine the quantity of dead leaves, branches, and twigs that this oldtimer has decomposed in its lifetime.

A major group of soil microbes is the Actinomycetes or ray fungi, a name that refers to the starlike growth pattern of some of its species. In addition to their role in releasing nutrients to the soil through the decomposition of plant and animal remains, they are the source of many antibiotics in common clinical use today. The organisms themselves often appear to be halfway between bacteria and fungi (some authors have called them Actinobacteria). They have funguslike filaments, but these may break up into individual cells and some are individual cells all the time. Apparently it is the Actinomycetes that give soil its "earthy" smell. Members of the order probably evolved antibiotics to suppress their bacterial and fungal rivals in the race to colonize food particles in the soil. We say "probably" because although the suppression of bacteria and fungi can be easily demonstrated on an agar plate in the laboratory, it is remarkably difficult to show the same process in the soil, where all kinds of organisms are simultaneously fighting for all kinds of resources.

Agricultural scientists constantly study soil microorganisms, but perhaps the most time and resources are used to explore a group of bacteria belonging to the family Rhizobiaceae, or root bacteria. Within this family, bacteria belonging to the group Rhizobium are not really soil dwellers, in that they invade the roots of plants and cause small swellings or nodules (Figure 14). Despite sounding like a serious threat to the plants, the invasion is exactly the opposite: the bacteria are able to absorb nitrogen from the air in the soil around the roots and transfer it to the plants. In other words, nature has evolved a built-in fertilizer factory for those plant species that can accommodate the bacteria. This process, technically called nitrogen-fixing, is an excellent example of symbiosis—that is, two species living together intimately. It is also a fine example of mutualism—that is, two species

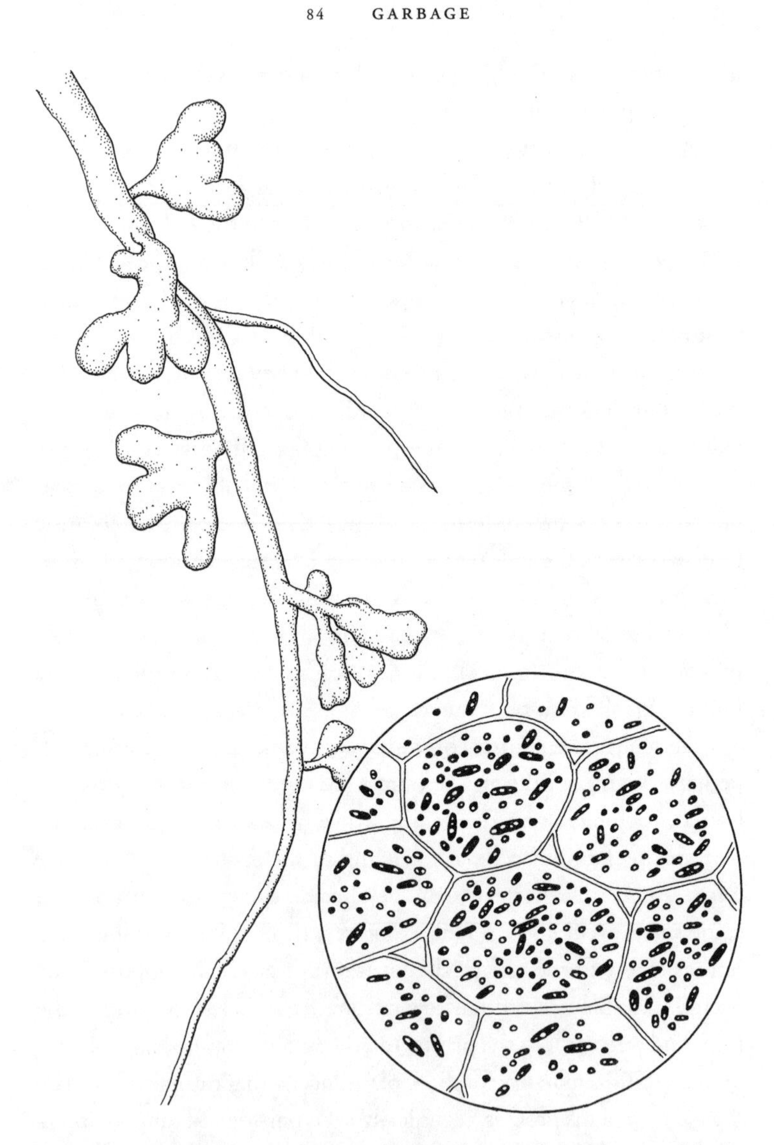

Fig. 14 Section of the root of a plant, showing the special growths or nodules in which nitrogen-fixing bacteria live. The enlarged portion depicts the interior of a nodule with root cells containing the bacteria.

that benefit each other (the plant is fertilized and the bacteria have a safe place to live).

Unfortunately for agriculture, relatively few plant families can accommodate nitrogen-fixing bacteria. To be sure, one of them, the legume family, is of major agricultural importance all over the world, providing the huge variety of peas and beans that are part of the diet of a great many different cultures. Yet staple crops such as rice, wheat, corn, and potatoes cannot accommodate nitrogen-fixers. Biotechnologists all over the world are therefore carefully examining every molecule involved in the bacteria-root symbiosis to see if, on the one hand, the bacteria can be persuaded to grow in strange roots and, on the other, if crop plants can be persuaded to accommodate strange bacteria. If you consider for a moment the number and variety of bacterial invasions that are harmful diseases, you can imagine the difficulty of coaxing a plant to allow its roots to be taken over by bacteria with which it has not evolved. Nature has provided the self-fertilizing model, the inspiration; now we humans are trying, with some success, to make use of it to boost the world's food supply. The global annual use of nitrogenous fertilizer is about 70 million tons, at a cost of roughly $150 per ton, so any success in taking advantage of this wild solution will generate major savings.

Bacteria are not the only microbes to be welcome in the roots of plants; there is an influential underground partnership between fungi and roots (Figure 15). The fungi are burdened with the hideous name mycorrhizae, a word derived from the Greek "myco," meaning fungus, and "rhizo," meaning root. These fungi invade the roots of most plant species. While taking up residence within or among the cells of the root, many of their filaments remain outside, in the soil. The fungi act just like root hairs, taking up water and nutrients from the soil—in effect providing the plant with a spare root system. More than five thousand species of fungi are involved. For many plant species from orchids to forest trees, the fungi are essential as, without their inva-

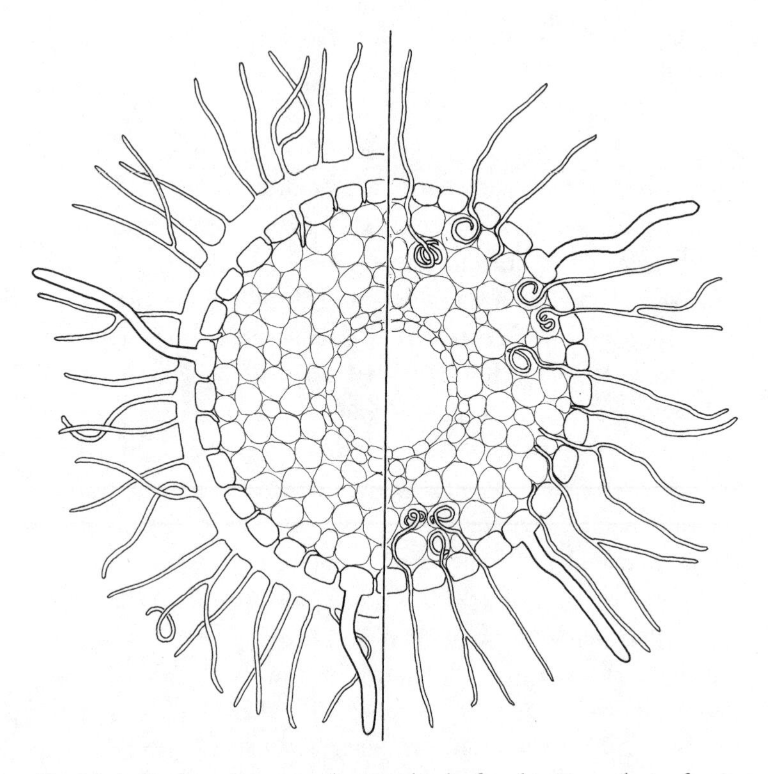

Fig. 15 A slice through a root to show two kinds of symbiotic, root-loving fungi. On the left is the type that forms a sheath around the outside of the root with only a few fungi penetrating the root itself; on the right is the type that inhabits both the interior and the exterior of the root. While this partnership can only be seen under a microscope, even the largest forest trees require it for healthy growth.

sions, the plants would grow poorly or die. In this subterranean world, the community can become complicated: more than one kind of fungus can be involved with a single plant species, and very often bacteria are present, some so useful that they are known as mycorrhization helper bacteria.

As you may imagine, a spare root system sounds like a dream

come true for farmers and foresters who are continually looking for ways to get better yields. Modern biotechnology is very interested in these fungi and bacteria for crop improvement. Unfortunately, the fungi are extremely difficult to grow in the laboratory, so progress has been slow. The immediate aim is to persuade crop plants to accept invasion by the fungi, so that the ability to absorb nutrients from the soil is increased. The ultimate goal is to reduce the use of artificial fertilizers, which are not only expensive but often pollute surface and ground waters. Another more subtle goal of the researchers is defense. Some kinds of soil fungi, unlike mycorrhizae, are extremely harmful invaders; once they penetrate the root, the plant becomes diseased and dies. The idea is to defend the plant against these killers by making sure that the areas they attack are already occupied by friendly fungi, leaving the harmful invaders no room for maneuver.

Soil Microinvertebrates

The second major group of soil decomposers are the microinvertebrates, which at 2 millimeters or less are so small that they generally require a precision microscope to be seen and studied. They include single-celled animals such as the common amoeba (Figure 16), which changes shape as its jellylike body glides over mud and stones. Other single-celled organisms (known as Protozoa) make up a kaleidoscope of animals with different barrel, bell, pear, and footprint shapes. Many have filaments that row or propel the owners through tissue-thin films of water that bathe the minute particles of rock and sand (and other bits of organic remains and geological fragments) that make up the soil. Careful laboratory studies have shown that an average amoeba must eat somewhere between a thousand and a hundred thousand bacteria every day just to stay alive. This is life at the start of the food chain!

The microinvertebrates also include tiny animals such as rotifers

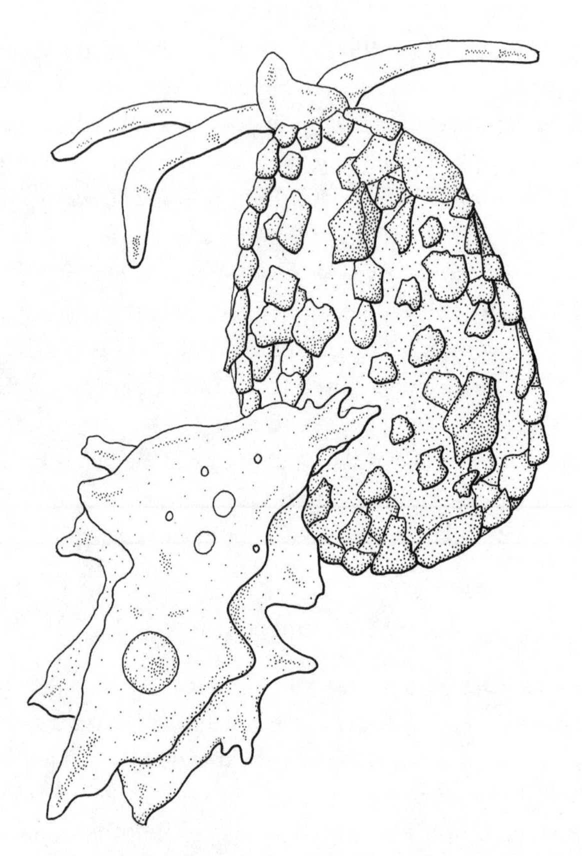

Fig. 16 Two soil amoebas, one with a jacket of soil particles, the other without. Single-celled animals such as these are major predators on bacteria.

and nematodes, both of which have a very large number of species a millimeter or less in length. Many species of rotifers are carnivorous and when greatly magnified have a monstrous appearance. Around the front end whirls a crown of closely packed hairs that give the impression of a rotating wheel (Figure 17). Within this crown are massive jaws capable of crushing prey. The entire animal may be covered

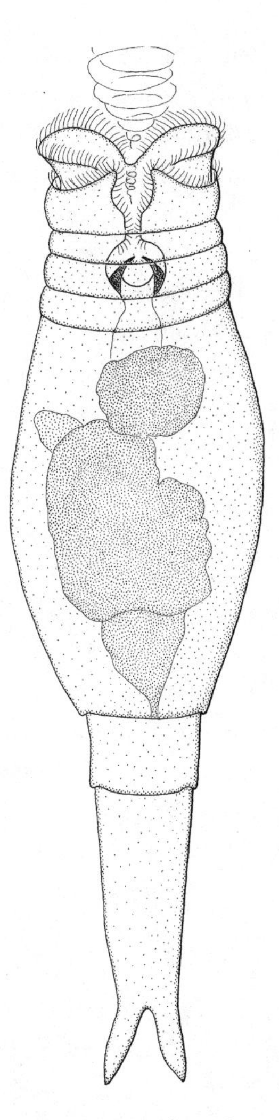

Fig. 17 A rotifer with its front end armed with hairs that waft tiny prey toward the jaws lurking just beneath.

by a lorica, a sheath of armor plates. Various spines and tentacles may be present as flotation devices or egg holders. The nematodes, by contrast, are transparent, threadlike, and rounded at either end—pretty dull at first glance. However, about eighty thousand species have been described, and another nine hundred thousand species may be awaiting discovery. Because they can reach densities of ten million to twenty million per square meter feeding on bacteria, fungi, or dead plant material, their effects on soil are profound. One nematologist pictured nematode numbers by imagining that if everything in a meadow except the nematodes were removed, a ghostly shadow made up of worm bodies would remain.

Soil Mesoinvertebrates and Macroinvertebrates

The mesoinvertebrates and macroinvertebrates are animals more than 2 millimeters long. They include a variety of mostly small species such as mites, springtails, and many nematodes, as well as more familiar groups that may be one to several centimeters in length such as earthworms, snails, beetles, ants, slaters, centipedes, millipedes, termites, and scorpions. Many soils contain several hundred thousand of them in every square meter. For example, small worms called enchytraeids, which eat bacteria, nematodes, and dead plant material, reach populations of two hundred thousand per square meter. Termites, which are much larger and eat all kinds of living and dead plant material including wood, can reach densities of several thousand per square meter. An observer examining a handful of soil will usually see a variety of worms and insects. The same handful under the microscope will reveal dozens more species, especially mites. In some soils mites increase to a quarter of a million individuals in the upper 10 centimeters of every square meter.

Some of these animals (scorpions, centipedes, and many ant and beetle species) are carnivores, but most are decomposers, consuming

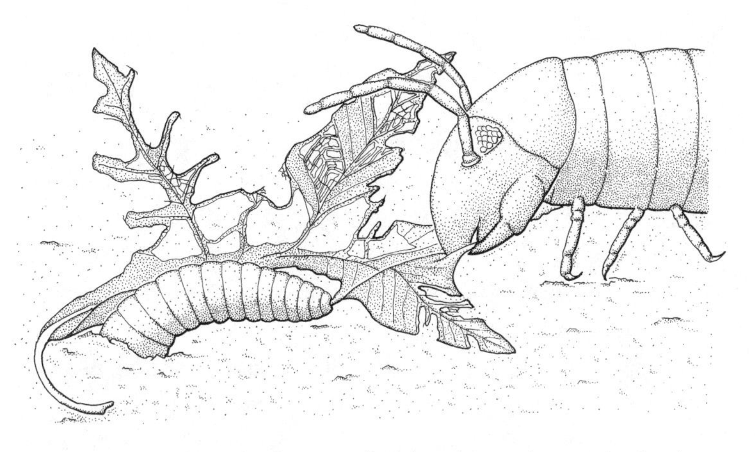

Fig. 18 Composting technology: A millipede and an earthworm shred and eat a fallen leaf. This process converts the valuable nutrients locked in the dead tissues into dung, an ideal substrate for the microbes that make the nutrients once again available for new growth.

dead material as either larvae or adults. Not all act in the same way. Millipedes and sow bugs are leaf-shredders, being large enough to munch on the softer material and often leaving the veins. Termites and beetle and moth larvae bore into logs and fallen branches, transforming wood into crumbs and powder. Earthworms haul leaves into their burrows below ground, and the armies of mites nibble any dead plant material, their vast numbers making up for their minute size (see Figure 18). Burying beetles inter corpses, and their larvae share the feast with those of many different kinds of flies. Numerous ant species nest in the soil, in many areas moving as much as earthworms.

The combined activities of these two kinds of animals have a major beneficial effect on soil fertility, aeration, and capacity to hold water. Whatever their activities may be, all of these organisms encourage decomposing bacteria and fungi in two significant ways. In the first, whole leaves or twigs or corpses are shredded into smaller pieces, creating huge surface areas for the microbes to settle on and attack. In the second, while the materials pass through the guts of the decomposers and much nutrition is removed to build their bodies, the dung nevertheless contains leftovers that the microbes can use. The result of this unplanned cooperation is that even a massive log, through the boring, scraping, chewing, nibbling, and digesting, bit by bit over the years, is gradually transformed from wood to powder and finally to simple molecules. It is these molecules that are the building blocks of new organisms (see Figure 19).

Now that we have outlined the world of natural decomposers, we can take a look at how they react to human activities and wastes, and how they not only provide irreplaceable services to us but also have become sources of new products and industries. We have already seen that certain bacteria and fungi that inhabit the roots of plants may one day be transferred to a range of crop plants and be a basic resource for agriculture. If this becomes feasible, the savings on the costs of both nitrogen and phosphorus fertilizer worldwide would be staggering—all based on tiny organisms from some obscure corner of the planet's biodiversity.

When we think of the mining industry, we conjure up massive images: haulage trucks the size of houses with wheels taller than the driver, giant excavators, drag lines, monster ore trains, huge tracts of country degraded by abandoned mines, artificial mountains of mine tailings, and lifeless waterways full of toxic wastes. By contrast, bacteria invite thoughts of microscopic objects at the extreme opposite end of the scale, objects so tiny that a mining engineer might inadvertently tread on several hundred billion of them every day! And why not, of what use are bacteria to mining? In fact, bacteria are becoming *so* useful to the industry that they are the foundation of a new way of extracting metals.

This revolutionary approach, known as biological mining, relies on finding species within the natural diversity of bacteria that can use

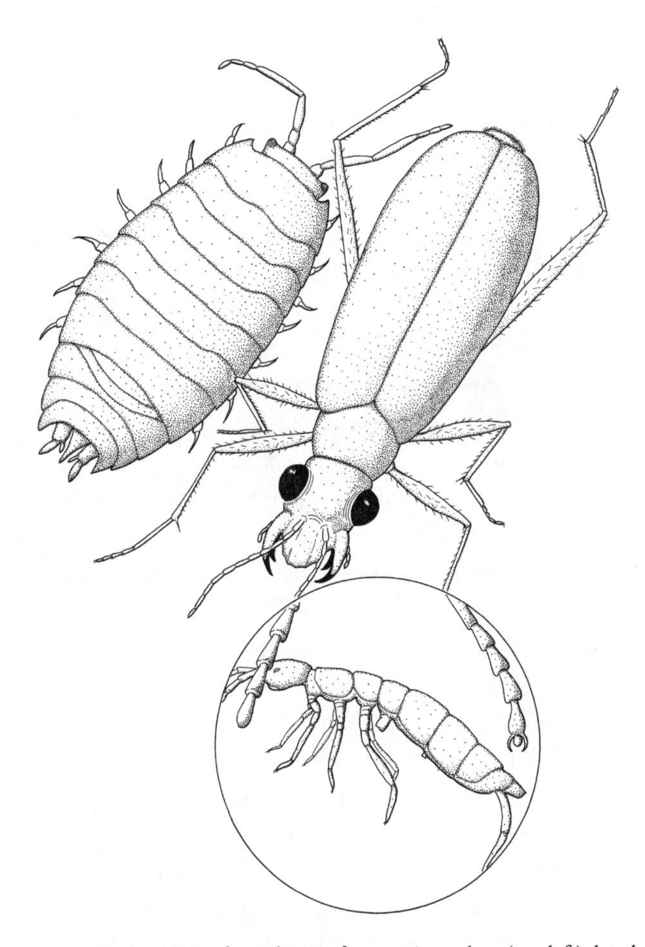

Fig. 19 Machine parts for soil manufacture: A sow bug (top left), beetle (center), and springtail (magnified). The sow bug and springtail eat dead and decaying materials such as leaves and twigs, and often occur in huge numbers. The beetle shown here is a predator, so one of its contributions is the conversion of prey to beetle dung, an ideal food for soil microbes.

metal ores and mine wastes as their "substrate" or food. However, biological mining has become attractive commercially for the pressing reason that high-grade metal ores are increasingly difficult to find. Even when located, they are often so deep in the ground or otherwise inaccessible that the costs of extraction are too high. Thus, effective methods of extracting metals from low-grade ores, or even from mine wastes, are at a premium—and that is where bacteria enter the picture.

Early experiments showed that a heap of low-grade copper ore, when sprayed with a mixture of a weak acid and copper-loving bacteria, produced a simple and easily recoverable form of the metal that drained into a collecting chamber at the base. This process could take months, but the addition of certain chemicals such as iron salts often speeded it up. In some circumstances the spray was not needed, as airborne bacteria found their own way to the test site. Thus, while the extraction was slow, it was inexpensive and turned a low-grade ore into a profitable end product.

Today, about a quarter of world copper production is obtained relatively cheaply by increasingly clever uses of bacteria. Biologists and engineers are screening thousands of different kinds of bacteria to find strains that will mine other valuable metals such as manganese, uranium, and gold. The search is becoming ever more sophisticated as scientists look for microbes that not only yield the desired metal, but do so quickly and under conditions where heat and lethal metals (such as mercury and cadmium) kill all other organisms. If the appropriate bacteria can be found, they will replace the enormously expensive engineering required to remove heat and toxic metals by the conventional extraction process.

Mine waste usually contains traces of metal and may be thought of as an extremely low-grade ore. With this concept in mind, it is easy to move from the new industry of biological mining to the even newer industry of biological remediation, or bioremediation. Bacteria are used to convert traces of metals, or any other pollutants, into harmless by-products, and the goal is to improve the environment rather than to extract for commercial gain. Nevertheless, with increasing public demand for the removal or destruction of pollutants such as metals, bioremediation is becoming a highly profitable business, particularly the cleaning up of derelict mines and factories.

Bioremediation often involves the removal of trace pollutants, that is, pollutants that may be present in very small amounts but are still sufficiently toxic to jeopardize health or to prevent the regeneration of natural vegetation. Often it also involves the removal of massive quantities of pollutants, and it is in this context that many people first encounter the industry. A dramatic introduction to bioremediation might be when an oil tanker runs aground and a catastrophic spill blackens the coastline. Bacteria are a major weapon in the fight to remove the oil. As in biological mining, exploration of the bacterial world has resulted in the discovery of extremely useful new species, in this case bacteria that use oil as their substrate and break it down into harmless components.

When disaster strikes, two principal kinds of treatment are available. In the first, mixtures or "cocktails" of bacteria are sprayed on the oil to break it down into components which they either use themselves or disperse harmlessly into the sea. Adding fertilizers, special nutrients required by the bacteria, usually speeds the process. Alternatively, the site is merely sprayed with fertilizers in the hope that oil-loving bacteria will be carried to the area by the sea or on the wind. In some cases, the natural abundance of bacteria has been so great that the oil has been densely colonized within a few days. Having arrived at the site, the bacteria find a banquet of oil, laced with sauces and vitamin supplements, courtesy of the bioremediation industry. Experts have noted two very interesting things about oil-slick bioremediation. First, the cleanup provided by naturally occurring bacteria may be far more effective, and is certainly much cheaper, than any man-made method. Second, oil-loving bacteria may be so common that it is probably very fortunate that oil reserves are deep underground; if they were not, some bacteria might have consumed them long ago!

Oil spills are common events; one estimate of the total amount of oil spilled into the oceans since 1976 is well over one billion liters. Thus, it has been a significant advance to recognize that the bacterial diversity of Earth can provide a solution to the problem. This diversity is a basic resource, a reservoir of ready-made, highly evolved technologies capable of helping clean up the environment. The search for more useful bacteria continues. Maybe some will be found where crude oil is part of the natural landscape. Perhaps a useful species will be found at places such as the La Brea tarpits in downtown Los Angeles, where tar bubbles to the surface underfoot as it has for millions of years—food for bacteria that regard it as home.

Recognition of bacterial resources is the basis of the bioremediation industry, which has expanded to include a variety of microorganisms including blue-green bacteria, yeasts, fungi, and single-celled algae. Cyanide is a well-known chemical used in gold extraction, and in many parts of the world it has become a widespread and deadly pollutant. Some Australian scientists conjectured that gold-mine wastes would be a logical place to look for bacteria that degrade cyanide, and they did indeed find them there. However, their discovery was ironic, for the miners regarded the bacteria as pests! The "sin" committed by the bacteria was that they removed the cyanide so efficiently that it could not be easily recovered for reuse; brand-new shipments were constantly required at great cost. Of course, outside of the mine-waste ponds, in the natural environment, the bacteria perform a valuable cleanup service.

Among the many useful bacteria harnessed for cleanup are those that feed on diesel spills in soils; cadmium, chromium, iron, and zinc in wastewater; arsenic in water supplies; car tires; waste sludges in oil fields and refineries; radioactively contaminated concrete in nuclear reactors; and discarded fuel additives. There are many more examples of the exploration of the natural bacterial world for wild solutions, among them for the ubiquitous pollutants known as PCBs (polychlorinated biphenyls). These compounds and their relatives contain chemicals such as chlorine, fluorine, bromine, or iodine; together they are known as halogens, and are notorious for two reasons. First, they are extremely resistant to breakdown. This makes sense, as the halogens are well-known antibacterial chemicals (for example, chlorine is added to swimming pools and iodine to wounds to keep them free of microbes). Second, they are now worldwide pollutants. The PCB family was used widely, especially in the manufacture of electrical appliances. When abandoned in waste dumps, PCBs have a remarkable ability to move around. The blubber of sperm whales stranded on European beaches has been found to contain these chemicals; since the whales feed chiefly on squid and fish, which they hunt in deep water well away from land, it is clear that PCBs are now in food chains far from where they are made. Polar bears contain high levels of PCBs, and some exhibit severe anatomical abnormalities that appear to be related to this form of pollution.

Bacteria that use PCBs as their substrate have been located precisely where one might imagine them to be—at derelict factories. These bacteria have been developed in the laboratory into cultures that break down PCBs. In one instance the breakdown products are a bright yellow, so that it is easy to tell when the PCBs are being destroyed.

In the search for organisms that would destroy halogen-containing compounds, bacteriologists have wondered if there are places where they occur naturally. Indeed there are. In some worms that inhabit the oozes and muds on the seabed, the desired organisms are by-products of the curious physiology required to live in such locales. The worms live in tubes, and the walls of the tubes contain halogen chemicals and—you guessed it—highly specialized bacteria. The bacteria have been brought into the laboratory with the hope that they will one day become useful bioremediation agents.

In a fascinating twist, some scientists have begun to look for agents that break down pesticides in drinking water. Where might such chemicals be found? One answer is in the guts of insects that have become pesticide resistant. Researchers have isolated the gene that confers the resistance and transferred it to an easily cultured bacterial species. This success has led to the possibility of filtering water through a bioreactor, a large vessel containing the bacteria that consume the pesticides. Although it is early in the development of this technology, the intriguing thought remains that insects' digestive tracts may be a valuable biological resource. Of course, the smartest approach would be to reduce our reliance on pesticides in the first place; it is always easier to prevent pollution than to clean up afterward.

When living in water or very moist environments, bacteria and other microorganisms frequently form a thin film, or biofilm, on every submerged surface. A well-developed biofilm may have several layers of bacteria, sometimes accompanied by algae and fungi, embedded in slime that sticks to the surface even in running water. The most familiar example is the plaque that forms on teeth, but biofilms are also found as slime on rocks or logs submerged in lakes and rivers, inside surfaces of aquariums (including the plants) and the piles of wharves and piers. They are also common in water pipes and other kinds of tubes that are constantly moist, such as the lining of our own intestines.

Biofilm can be very helpful, especially for water treatment. The trick is to expose the water to as much biofilm as possible and to create conditions that ensure that the biofilm contains useful bacteria. Treatment plants often incorporate rock that has been crushed, to expose many surfaces on which a biofilm can grow. Wastewater flows over these surfaces, causing the bacteria to either absorb pollutants or break them down into harmless constituents. Water engineers have discovered that it is sometimes the membrane on the outside of each bacterium that is the most active component, for the pollutant binds or sticks to it. The biofilm does not have to be alive, as long as the water is exposed to large surface areas of bacterial membranes. Using this sort of technique, bacterial engineering helps us treat the effluent from mines, factories, and nuclear power plants, removing pollutants such as zinc, copper, iron, manganese, lead, and uranium. Bacteria are also the mainstay of the treatment of that ubiquitous human product, sewage.

Bacteria are not the only biological miners or remediators. Botanists have known for decades that many species of plants are adapted to grow in soils that naturally contain high levels of metals such as copper, zinc, lead, and nickel. Adaptations in these metal-tolerant plant species result in storage of the metal in parts of the plant where it is harmless. For example, it may be stored in the cell walls in a form that does not poison the contents of the cell. Sometimes the agent is a special protein that binds (chelates) the metal into an inactive molecule. The plant species that have evolved in naturally contaminated soils are being explored for their mining and remedial potential. The simple observation that some plant species colonize and grow on mine wastes has suggested novel biological resources—plants that accumulate metals and may therefore be candidates for mining or bioremediation.

The search for plant species that absorb valuable metals such as gold in commercially important amounts has recently focused on a relative of the cabbage. Not very exotic perhaps, but under certain conditions this plant is a hyperaccumulator, that is, it accumulates gold in its tissues far above normal levels. The research continues, to determine whether plants such as this cabbage relative can be made to accumulate metals sufficiently to be of interest to the mining industry. In another direction, metal-tolerant plants can be useful in mineral exploration. A map showing the distribution of a plant species known to be adapted to a particular metal might be at least a rough guide to locations where the metal is near the surface. In some cases, these maps have shown engineers where to start drilling.

Many metal-tolerant and metal-accumulating plant species have been discovered around the world. They may be small in stature like the alpine pennycress and the zinc violet, which accumulate zinc, or they may be trees like the blue sap tree, which tolerates and accumulates nickel.

Even lowly lichens that accumulate uranium are showing promise. Many have potential for bioremediation, especially for the revegetation of abandoned mines and landscapes polluted by heavy metals. One technology under investigation is the "cultivation" of "crops" of hyperaccumulators in contaminated areas until most of the polluting metal has been extracted by the plants and stored in their tissues. These plants may then be incinerated, removed to a safe site, or perhaps even treated as a low-grade ore with the metal being recovered from them. Genetic engineering may enhance the ability of plants such as these to accumulate and store highly toxic wastes. Some genes produce enzymes that convert pollutants such as heavy metal salts into relatively less harmful forms. Successful transfer into hyperaccumulators opens the possibility of both "harvesting" the pollutants and breaking them down, all in a single plant.

Floating water plants such as the duckweed and the water hyacinth (see Figure 20), which have dangling roots, also show promise as "crops" for the bioremediation of waterways. Often regarded as weeds in areas where they are not wanted because of their rapid rate of growth, this very property can be applied to the removal of aquatic pollutants. Specifically, the duckweed absorbs dissolved fertilizer in the runoff from farmers' fields, and the roots of the water hyacinth absorb many kinds of toxic chemicals that flow out of mine wastes. Because of the floating life style of the plants, they can relatively easily be harvested from boats and either recycled to utilize the captured fertilizer (in the case of the duckweed) or safely disposed of (the contaminated water hyacinth). Once the plants have been removed, the surface may be quickly recolonized to prevent a new buildup of the pollutants.

While the future prospects for phytomining remain uncertain,

Fig. 20 The water hyacinth is a terrible pest where it has been introduced outside its home range, for it covers lakes and rivers with impenetrable vegetation. However, in some situations it helps clean polluted water by absorbing unwanted chemicals. The hyacinth is then harvested and burned. In some instances the ash is safe for use as fertilizer.

phytoremediation is already widespread. The examples discussed here illustrate that many different plant species have emerged as being of proven or potential commercial use. One needs to remember, though, that many hyperaccumulator species are restricted to very small areas where they have adapted to the local conditions and are vulnerable to extinction. Metal-tolerant or hyperaccumulator plant species are a biological resource worthy of protection. Who can guess what little-known or lowly plants may be badly needed in the future?

Finally, one of the most interesting resources to have emerged from the multitude of soil invertebrates is the dung beetle. Its value has perhaps been most appreciated by the Europeans who settled Australia. The cattle they took with them were the first to be seen on that continent. The indigenous Australians did not need cattle; they were hunter-gatherers whose culture had matured without herding domesticated animals over a period of at least forty thousand years, longer than anywhere else on Earth. Nevertheless, the settlers soon were breeding massive herds that roamed throughout much of the country. While the numbers fluctuated wildly through years of high rainfall or drought, one unexpected and persistent problem foreshadowed the nightmare depicted early in this chapter: what if there were no decomposers?

The situation that developed was as follows. Australia, being a large island, has been more or less isolated from the rest of the world for millions of years. Neither cattle nor any of the decomposers that might consume their dung had ever evolved there. Settlers brought the cattle but no decomposers, so as the cattle population increased, the dung began to accumulate in alarming amounts. At one time it was estimated that there were 30 million cattle producing 300 million cowpats each day, covering at least 2.5 million hectares of pasture each year. Australian scientists discovered that a variety of dung beetles adapted to the native marsupials had evolved Down Under. However, no species preferred cattle dung, which was not surprising as these animals were aliens to the land. The researchers went on to ask the kinds of questions we have been posing in this book: Were there beetles that ate cattle dung? If so, where had they evolved?

Cattle were domesticated in several areas of the Northern Hemisphere and were prominent in the early history of various African cultures. Accordingly, the search for dung beetles focused on the continent of Africa. The beetles were found and brought to Australia, where several species from both the northern and southern extremities of Africa turned out to be major cowpat consumers. The African beetles were cultured in the lab and set free in the areas most afflicted.

The benefits appeared rapidly. The beetles broke up the cowpats, tamping them into dung balls with their hard, shiny bodies, then buried them and laid eggs in them so that the larvae would grow up surrounded by food. The disappearance of dung from the landscape was not the only benefit. Much more important was the fact that it was buried and transferred the fertilizer content into the soil, where it was beneficial. The beetles' digging activities also yielded healthier roots and better water retention in dry soils. Up to this point, dung dissolved with the first hard rain that washed it into streams and rivers.

Removal of the dung had one final, very welcome spin-off: the dung was home not only to beetle larvae, but also to the larvae of many different kinds of flies—which, unlike the beetles, quickly adapted to the presence of this banquet of foreign food. Before the introduction of beetles, the flies had a near-monopoly in the cow dung; their larvae multiplied without competition from beetle grubs and hatched into clouds of adults that became a serious nuisance to man and beast alike. Once the beetles came and the fly larvae had to share their food, the fly menace subsided.

Today, companies breed dung beetles to order, the latest usage being public parks in cities where negligent dog owners have allowed their pets to foul the grass. The beetles approach their meal from below, and the smelly piles sometimes disappear as rapidly as overnight. Companies that have searched biodiversity for useful products know that dung beetles have a high dollar value. Exploration has allowed breeders to find beetles for particular climates and local conditions. How intriguing that the same group of beetles were highly prized by ancient Egyptian civilizations and were even regarded as sacred, a symbol of resurrection and immortality! Yes, dung beetles are scarabs. Perhaps there is a connection with resurrection, symbolized, in turn, by the decomposition and recycling that the beetles promote.

Dung beetles provide only one example of the way invertebrates supply ecosystem services. Among other soil invertebrates, perhaps the best known and most admired are earthworms. Farmers and gardeners have always appreciated the worms' activity: they drag leaves below ground, pass soil through their digestive tracts, and forge miles of burrows, fertilizing and aerating soils. Worm casts are higher in nutrients such as nitrogen and also contain more bacteria than the soil that surrounds them. Scientists have performed careful experiments to see if crops and garden plants grow better when earthworms are present than when they are not. Indeed they do, and many farmers foster worm growth to save precious dollars otherwise spent on fertilizers. The produce of organic and integrated farms, where the activity of soil invertebrates is encouraged, is increasingly sought out by the public, as the products tend to have been grown with lower levels of fertilizers and pesticides, or none at all. It is ironic that in many parts of the world where farmers cannot afford agricultural chemicals, reliance on soil invertebrates such as earthworms has always been well understood.

When it comes to recognition of the value of soil invertebrates, the earthworm is just one example. As we have seen, every square meter of a healthy soil contains thousands of decomposers that interact with millions of microbes. It is odd that scientists have not focused on these decomposer armies to demonstrate exactly what they do, and how many species it takes to do it. A few experiments have shown that the addition to leaf mulch or soils of different kinds of invertebrates such as beetles, springtails, or termites can definitively accelerate decomposition. Conversely, if these organisms are removed, recycling declines and plants grow more slowly.

We humans remain ignorant of much that happens in the soil. Consider the student who examined the soil mites in several forests in eastern Australia. As mites are so tiny, his samples were confined to small cores, each about the diameter of one's middle finger, to a depth of 10 centimeters. Although he took many such cores, the total area sampled was less than a square meter. Even in this small patch he found 172 species! What they all do and whether or not all are needed to maintain the forest soil is unknown. The same sort of situation holds for most of the world's soils, the soils we depend on for agriculture and forestry. Put another way, soil is a vast, complex machine that generates necessities such as food, and commodities such as timber; but we have little idea how many parts constitute this machine, or what they do. Our prediction is that increased knowledge of their resource value will generate revolutions in industries such as agriculture and forestry.

As we close this chapter we cannot resist talking about the role of invertebrates in crime detection. As we have seen, corpses are valuable food for a variety of decomposers that appear to share the feast in a remarkably civilized way. For example, a dead sheep is first colonized by blowflies, but as changes are brought about by the activity of their larvae and by bacteria, other kinds of flies, beetles, and moths take their turn at the slowly vanishing remains. Experiments in which dead animals have been placed in various situations show that "waves" of different insects follow each other in the corpse in a process known as succession. The species involved in each wave, and the time each wave lasts, depend on a variety of factors such as temperature and whether or not the corpse was buried or the animal had drowned.

The horrible truth is that we know a lot about succession on human corpses—enough, in fact, to pinpoint the time of death by the decomposers inhabiting the cadaver. In several cases, the alibi of a suspect has been destroyed and conviction for murder has followed. Conversely, similar evidence has been used to clear the wrongly accused. A Hungarian ferry skipper was convicted of the murder of a passenger after he started work at six o'clock one September evening. The decomposers inhabiting the corpse were studied at the time, but the information was not used in the trial. Eight years later, when the case was reopened, an entomologist showed that the three species of flies found in the corpse never fly in the evening; further, the state of development of their eggs and larvae in the corpse, given the cool September weather, was such that the murder must have taken place long before 6:00 P.M. that day. Since the captain had an iron-clad alibi prior to coming to work, he was released from prison.

That skipper had an immediate reason to be thankful for maggots. We *all* have reason to be thankful for maggots and for the other decomposers around the world that dispose of wastes and garbage for us.

SEVEN

Natural Enemies Are Best Friends

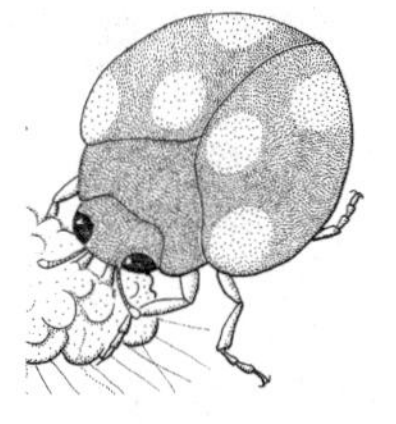

One of the least appreciated ecosystem services that biodiversity supplies to us is the control of agricultural pests by natural enemies. This was brought home to biologists some time ago by what is known as the Cañete Valley cotton disaster. Cotton became the principal crop in the Cañete Valley of Peru (south of Lima), replacing sugarcane and other crops, when irrigation was expanded in the 1930s and 1940s. After World War II entomologists advised growers to use cultural control measures to protect the cotton by, for example, destroying weeds that harbored the insect pests that attacked cotton plants. But the lure of new synthetic pesticides was too much for the growers. DDT and several of its "chlorinated hydrocarbon" chemical relatives were widely used in the valley starting in 1949. At first it seemed the growers had chosen wisely: cotton yields that had been at 495 kilograms per hectare in 1950 rose to 729 kilograms per hectare in 1954. As yields climbed, growers concluded that the more pesticides they used, the better off they would be. The Cañete Valley was blanketed with them, and many trees were chopped down to make it safer for crop-dusting aircraft to spray the fields.

The paradise was short-lived. By 1952 cotton aphids had become resistant to one DDT relative, and by 1954 the tobacco hornworm (a moth caterpillar) was resistant to another. By 1955–56 boll-weevil infestations were severe, and another moth-caterpillar pest exploded to unprecedented populations and was highly resistant to DDT. In addition, at least six new pests appeared, ones that were not problems in other valleys which had not been subjected to a synthetic pesticide blitzkrieg. This situation was a result of the vulnerability of predatory insects to the pesticides. Those predators, which normally kept populations of cotton-eating insects at such a low level that they were not considered pests, were decimated by the pesticides. The populations of obscure cotton-eaters, freed from the assaults of predators, increased in size and were "promoted" to pest status. Despite the application of other pesticides, cotton yields plummeted to 330 kilograms per hectare. Only then was integrated pest management instituted via programs that took advantage of the presence of natural enemies of the pests. By 1963 yields had climbed to over 780 kilograms per hectare.

One would think that the lessons of the Cañete Valley would have been taken to heart by farmers (or at least cotton growers) the world over, but alas they were not. In the next decades, insecticide applications to Central American cotton fields increased from about ten per season to forty! The evolution of resistance led to new outbreaks of old pests, and the decimation of natural enemies led to promotion of additional insects to pest status. Many farmers went bankrupt. In northeastern Mexico, pesticides used against the boll weevil clobbered the enemies of the tobacco budworm, which was then promoted to a pest so serious that 300,000 hectares planted to cotton in the 1960s were reduced to 500 hectares in 1970. Many of those employed in the cotton fields could not find substitute work and were forced to relocate.

The lessons of the Cañete Valley were not fully absorbed by the agricultural community, and serious problems from overuse of pesticides persist. One of the principal threats to the production of rice, the world's single most important crop in terms of the number of people it feeds, is an insect known as the brown rice planthopper. Virtually every outbreak of this pest in tropical rice-cultivation systems has been a result of intensive and widespread use of synthetic pesticides.

The chemicals attack not just the planthopper, but organisms such as spiders, crickets, and small predaceous bugs that prey on the planthoppers. In general, the spraying perturbs the agricultural system in a way that favors herbivores such as the planthopper. It has been estimated that use of the insecticide deltamethrin causes increases of about 1.5 million insect herbivores per hectare because of the differential impact of the poison on the smaller, less-resistant populations of their natural enemies.

The delivery of natural pest control is usually subtle and very much "behind the scenes." One is almost never aware of the service until it is disrupted. This principle was brought home to us on a long-ago field trip to Queensland in Australia with one of the world's leading ecologists. Charles Birch was looking for fruit flies for his research, whereas we were interested in finding caterpillars of a small moth that feeds on *Opuntia* cactus (prickly pear). Although we saw occasional clumps of the cactus, none showed the damage caused by the caterpillars and we never found the moth *Cactoblastis cactorum.* We were bewildered by this mystery. What was happening?

Understanding the answer to that question involves a little history. Early settlers in Australia imported *Opuntia* as an ornamental plant. The cactus escaped cultivation and, finding a physically suitable environment in which it had no natural enemies, spread rapidly. By 1925 thick growths of *Opuntia* covered nearly 256,000 square kilometers of New South Wales and Queensland on the continent's east coast. Over half of that area, the plants were so dense that the land was completely useless. Furthermore, the costs of clearing the land mechanically or of poisoning the *Opuntia* were greater than the value of the land itself. Australian biologists went to South America in search of natural enemies of *Opuntia* and eventually discovered that *Cactoblastis* was a potent foe of the prickly pear. The moth was introduced and with amazing rapidity gobbled the cactus until it was reduced to a few scattered clumps (see Figure 21). And that is how it has remained,

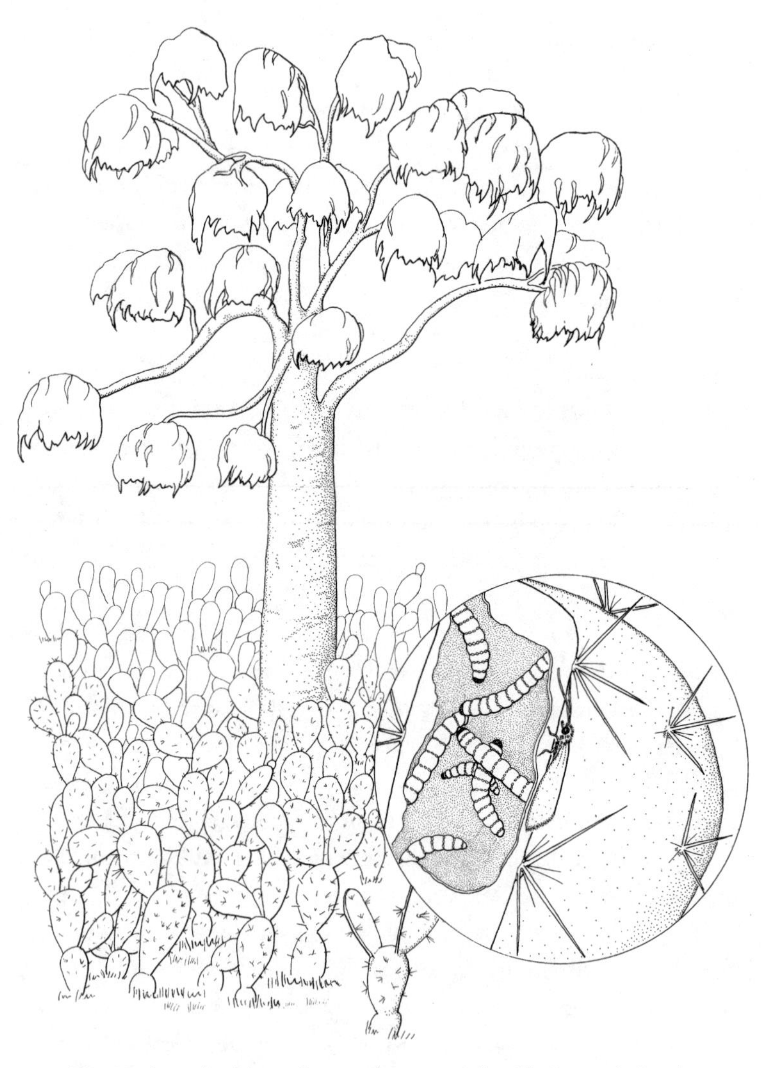

Fig. 21 A patch of Australian eucalyptus woodland before and after the introduction of the cactus-eating moth. At one time cactus covered the ground, outcompeting all native plants except trees. The figure on the left shows the

caterpillars magnified, eating a cactus leaf, while an adult moth awaits a mate. The illustration on the right shows a young tree fern, some native irises and grasses, and a few palm seedlings amid the tattered remains of the cactus.

as what is known technically as a "fugitive species." A clump of cactus will thrive somewhere until discovered by a female moth. Then it is doomed to be converted into many more moths, which disperse in search of other *Opuntia* as the clump is devoured.

Years ago scientists did not realize what powerful enemies of plants insects could be, but the *Opuntia* story helped to persuade them. The irony is that if the history of the cactus-caterpillar battle were not known, no one would guess that the rarity of prickly pears in northern New South Wales and Queensland is due entirely to the presence of a tiny moth, so cryptic that none of us could find it!

Since the supreme importance of natural pest control has been recognized, there have been more than 5,500 attempts to initiate it. Researchers have introduced specific insect predators and parasites from different areas in attempts to suppress populations of insect pests and weeds, a process known as biological control. The first such introduction, in California in 1888, was an attempt to control a destructive insect pest of citrus trees called the cottony cushion scale. A predaceous ladybird beetle and other parasites from Australia were brought in; they multiplied, spread, and did an effective job of controlling the scale. Although many introductions have been unsuccessful, about 165 insect pest species and 35 weed species have been adequately controlled in one or more countries.

Not all so-called biological control programs have been directed at insect pests and weeds; sometimes microorganisms are used. For instance, European rabbits arrived in Australia in 1788 and some years later were encouraged by "acclimatization societies" among the European colonists to invade the countryside. Such societies, made up of Europeans who wished to be surrounded by familiar animals from their homeland, were astonishingly successful with the rabbits, starting in 1859. Rabbits were also transported extensively as a source of meat for people. Lean wild versions of the rabbit that were introduced took to the countryside much more readily than the ungainly domesticated strains; unfortunately, they "bred like rabbits." Soon billions of them began to outcompete sheep for Australia's valuable grasses, and the sheep had already overgrazed much of their range. The rabbits not only ate grass, they also consumed the roots and bark of shrubs and small trees. Shooting, poisoning, and other control methods were of no avail. Thousands of kilometers of fencing were built to keep the rabbits out of unoccupied areas; one fence alone stretched more than 1,500 kilometers. But many were built too late and it was impossible to maintain them to the required level of perfection. The rabbits continued to spread. Cattle and sheep died of hunger; ranches went bankrupt and the owners had to abandon their land. It was worse than the *Opuntia* disaster.

The organism that rescued Australians from the European rabbit was even smaller than the *Cactoblastis.* In this instance the warp and woof of the natural internet were partially restored by the myxoma virus, which was successfully introduced into Australia in the early 1950s. This viral wild solution was found as a parasite of rabbits in the Western Hemisphere, where it infects a substantial fraction of the population but produces only mild symptoms. When the same virus invades European rabbits, however, it causes a disease that quickly kills them. In Australia the virus was carried from rabbit to rabbit primarily by mosquitoes, and at first the control program was extremely successful. Over 99 percent of the rabbits in infected populations died.

But then natural selection took over: the quicker a rabbit died, the smaller the chance that some of the viruses in its blood would be picked up by a hungry mosquito. Simultaneously, the rabbits responded to the virus by becoming more resistant. As a result of the evolution in both host and parasite, rabbit populations began to recover and quickly increased. In response, another virus has been released and the rabbit populations have again been decimated. While the impact of the virus has been immense, the familiar signs of resistance have been detected, so the battle continues.

Acclimatization societies have been the instigators of many ecological catastrophes. In some North American locations the European starling has grown to plague proportions because people wanted to have in their yards every bird mentioned by Shakespeare. No handy virus has been found that will control this bird, which competes for resources (including nest holes) with many native species. The lesson of the starling and many other creatures is clear: moving organisms around is one of the actions that causes most damage to the natural internet. For we are often shifting them away from their natural enemies, enemies that are our friends.

Biological control therefore carries risks. Today we have strict procedures and guidelines on the introduction of species for biological control. For example, before a herbivore such as a beetle is introduced for pest control, populations are cultured in strictly quarantined laboratories and tested to see if they will eat native plant species. With any hint that they may destroy local plants, they cease to be candidates for biological control. These careful procedures were elaborated after a series of disastrous introductions. Mongooses injected into the environment to control rats generally ignored them and wiped out native birds and lizards that were easier to catch. Toads, introduced into Australia to control pests in sugarcane fields, snacked on the pests but feasted on any local animal species they could catch.

Even when tests are made, it is not always possible to predict the eventual outcome, as the case of the flowerhead weevils demonstrates. These beetles, native to Europe, were released in North America for the biological control of *Carduus* thistles, which had been introduced from Europe and Asia and had become major weeds. Tests before the release suggested that the weevils prefer laying their eggs on *Carduus,* whose flower heads they eat, and that this would limit the degree to which the weevils would attack native American thistles in different genera. Unfortunately, the beetles spread geographically and began to attack native thistles. Since the weevils also happily eat the exotic thistles, they pose a special threat: when their attacks reduce the populations of native thistles, the weevils need not decline concurrently because they can utilize the nonnative alternative food.

Furthermore, the weevils are having a negative impact on native picture-winged flies, which share the thistle-head food resource. The relationships between plants and their herbivores and among different herbivores are indeed complex. A thorough understanding of those interactions is crucial before biological control agents are released, as is an acceptance of the great value of natural enemies, which only rarely present troublesome problems.

One way of avoiding the hazards of introductions is to place greater reliance on natural communities. In recent years this has been done in various parts of the world, generally amid dire predictions of failure from companies that manufacture chemical pesticides. One of the most dramatic success stories comes from Indonesia, where the old enemy, the brown rice leafhopper, became a vexing pest in spite of steady increases in pesticide application. The chemicals caused major declines in the pest's natural enemies, and as the pest increased in numbers it increased in resistance to a variety of pesticides. Eventually so much pesticide was in the air that many country folk became sick; in fact, pesticide poisoning was like an epidemic disease. The result was a total ban on the worst of the pesticides. Economists were horrified, predicting massive declines in rice yields and collapse of the local agriculture. Indeed, for several years yields did decline, but not in dramatic fashion. Within a few additional years, yields returned to nearly the levels obtained when pesticides were used.

What happened was that the local predators that survived the pesticides (spiders, ants, beetles) discovered a bonanza of brown rice leafhoppers and set to. There was a pause in the attacks on the pests as the predators, fattened on regular meals of leafhoppers, laid extra-large batches of eggs and reared more young than usual. The growing army returned to the rice fields, predation started again, and over the next few years the natural community of leafhopper predators built up to a point where pesticides were redundant. It is interesting to speculate on the worth of these anonymous spiders, ants, and beetles. It is hard to estimate the cost of the pesticides that had been used, but it must have been hundreds of millions of dollars each year. The local predators, although each one was small and insignificant, together were worth serious money.

Local communities or assemblages of natural enemies including predators, parasites, and pathogens (killer microorganisms) have provided crucial economic and environmental advantages in many countries of North and South America, Europe, Asia, and Australia. For instance, spring wheat in some parts of Idaho is protected from aphids by a mix of local insect-attacking fungi and minute parasitic wasps. The scientific name of one of the fungi is *Pandora,* but its box of tricks appears to be beneficial in contrast with the box opened by the princess in Greek mythology. Experiments are in progress to identify ways in which *Pandora* can be encouraged in crop plants. In Europe it has been shown that hedgerows and fallow fields are important sources of biological control agents, while in the tropics, strips of forest that permeate plantations provide the same function. The significant principle here is that varied landscapes provide more natural pest control than crop monocultures. Mixtures of crops and natural vegetation supply more varied habitats for beneficial organisms. This conclusion will produce a broad smile from old-time farmers the world over who have always recognized its validity.

A word of caution: almost certainly there will always be a need for pesticides, for the simple reason that variation among natural communities is normal and the right insects may not be around in sufficient numbers when they are needed. In such circumstances, pests may have to be controlled chemically. The great advances in pest control have arisen from the recognition that control can be achieved by biological as well as chemical means. Usually both are needed, and the science of integrated pest management (IPM) has evolved to discover the balance of biological methods required to maintain crop yields, human health, and environmental health all at the same time.

The need to conserve as many species as possible has been clearly demonstrated by the biological control industry. The examples discussed so far highlight a variety of obscure species that have suddenly attained commercial and industrial importance because they attack other species that we consider pests. Who would have guessed that a tiny, cactus-eating caterpillar from South America would one day save agriculture across a vast region of Australia? Right now there is some traffic in the other direction as scientists look for insects in Australia that might eat melaleuca trees. These were introduced into southern Florida, where they have become a plague in the absence of their normal Australian enemies. The present focus is on one species of small weevil, a kind of beetle with a sluglike larva that relishes melaleuca leaves. The story of the brown rice leafhopper shows that we never know which species are going to become pests as agriculture progressively modifies the environment. The message is that species should be conserved; it is impossible to anticipate which will turn out to be vitally important in the future. In this context, let us review briefly the kinds of species that are being marshaled to help control agricultural pests.

Flies seem like pests to us but they can be extremely useful. Species that attack plants in greenhouses can be controlled by releasing predatory flies that capture their prey on the wing, transporting them to perches where they suck the victims' juices. Other species are being recruited to tackle the huge problem of the fire ant invasion of the southern United States. This ant species was introduced from South America, and exploration of its native habitat revealed the presence of the scuttle fly that hovers above the ants, dive-bombing them to lay eggs on their heads. When the larvae hatch, they eat parts of the ants and kill them. One possible special advantage of this fly as a biological control agent is that its presence clearly bothers the ants, which constantly adopt defensive behavior. With the edge taken off their aggression in this way, local ants may be able to regroup and provide resistance on the ground. Further, the ants often cower under rocks and twigs, waiting for the flies to move away, an action that may seriously disrupt their foraging and hence diminish the food supply to the colony.

Various mite species are pests in crops, orchards, and greenhouses but fortunately many are highly vulnerable to other mite species that are predators. Most predation in this world is not of the kind commonly seen in nature documentaries—lions killing zebras or bears crunching salmon—but occurs instead at the microscopic level. One of the most common encounters is the attack and devouring of one mite species by another. A large-scale industry has developed around this interaction, and in many parts of the world huge sheds harbor billions of pest mites, lovingly grown to feed predator mites for release into the field. Needless to say, the pest factories are mite-proof so that they do not contribute to the problems in the field. Predatory mites are in use in situations as far apart as apple orchards in Australia and cassava fields in Africa.

Ants have frequently been considered to be biological control agents, and they are certainly important in natural communities that contribute to pest control. However, when individual species are developed as control agents, there may be difficulties in controlling the ants! They can be extremely effective predators (for example, searching out millions of plague caterpillars in Canadian forests), but they often have habits that are counterproductive. One of these is the use of aphids and their relatives as a kind of analogue of cows or sheep. The aphids suck the sap of plants but rarely use all that enters their digestive system. The excess is known as honeydew—a curious name for the sticky fluid that emerges from the back end of the insect! Many ant species relish honeydew and herd the insects as though they were domesticated, protecting them from their enemies and taking advantage of the nutritious fluid. The end result can be an infestation of aphids that may drain the plants, an unfortunate complication of a system meant to control pests. Experiments to find ant species without bad habits continue. As many thousands are available, we are confident that these superabundant and highly persistent predators will one day become useful biological control agents.

The plant *Echium plantagineum* is a pretty wildflower in Europe, but upon introduction into Australia it has become a weed, raging across hundreds of thousands of square kilometers of grazing land. It turns the landscape purple with its flowers and costs farmers more than $250 million a year in control methods and lost production. Not only does it displace the local forage grasses, but it is poisonous to stock and has become known as Paterson's Curse. Who Paterson was, we can only guess. The principal search for biological control agents has taken place in southern Europe. It appears that there may be two agents that can be released together. Extensive trials have been conducted with a weevil species whose larvae eat the roots and whose adults eat the leaves, and with a moth whose caterpillar also attacks the leaves. The results are promising.

Several species of water plants undergo population explosions when they are introduced into parts of the world where they have no natural enemies. The water hyacinth can be a useful agent of bioremediation when strictly controlled, but both it and the fern *Salvinia* float on the surface, often growing so densely that they clog rivers, canals, and lakes. They block all light to the water, kill the underwater plants, and cause the food chains to break down. Grave problems result in countries as far apart as Uganda and India, where fisheries are destroyed and boats are unable to move. In Australia, *Salvinia* is especially threatening because surface water is precious in that dry continent. Scientists went to the fern's native South America to find a biological control agent. The most promising was a weevil, but its release into Australian waterways was a failure. However, a second species of weevil that looked almost identical to the first proved to be an extremely effective control agent. A little later, in Papua New Guinea, the same weevil species reduced *Salvinia* cover from 250 square kilometers to 2, destroying two million tons of the weed in just two years. Clearly, it is impossible to know when a particular species is going to be of vital importance. When you've seen one weevil, you haven't seen them all! Conservation of species in countries other than one's own can be crucial.

We have elsewhere discussed the insects called parasitoids. The value of these animals to global food production is immense, as they provide vast armies of larvae that consume pest species. About 10 percent of all insect species are parasitoids, most of them tiny solitary wasps, but a considerable number are flies and beetles. They have evolved all sorts of ways of getting into their victims. Sometimes the adults lay their eggs on the surface of the prey; the larvae hatch and bore their way inside. Other species lay their eggs close to their victims, and it is the larvae that find and penetrate them. Still other species cleverly arrange to lay their eggs on the victim's food, so that the larvae have a free pass into the digestive system. In some cases, the wounds caused by insects attacking crop plants such as corn and tomatoes release chemicals into the air that attract parasitoids. These chemicals are being synthesized and sprayed on crops to see if farmers can recruit parasitoids on a regular basis; the results look favorable. The next time you see one tiny insect hovering near another, take a closer look; it may be a parasitoid seeking a home for its children.

We have visited factories where parasitoids are grown for release into fields and orchards. In one, shelves from floor to ceiling are stacked with squashes, each of which is covered with the pests known as scale insects (see Figure 22). Look closely and you see a faint haze made up of thousands of tiny wasps laying eggs on their victims. Return in a week and the haze has become a thick cloud, as the youngsters join

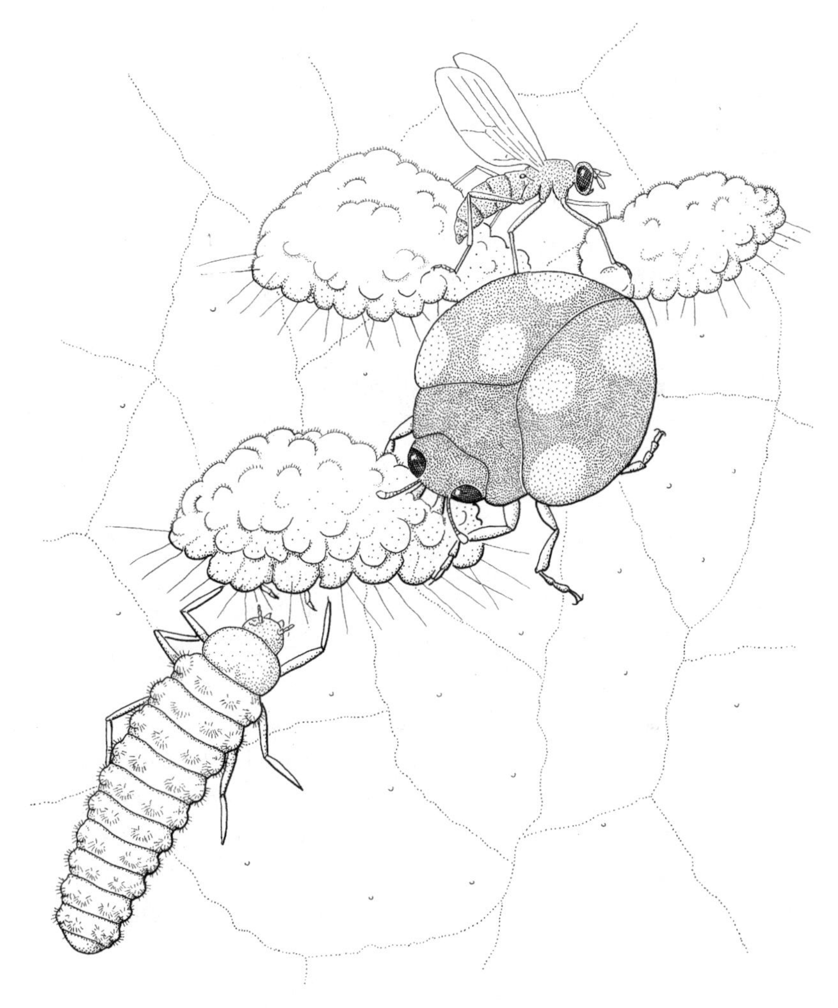

Fig. 22 Hard times for a scale insect on the leaf of a crop plant. A ladybug munches on one end while its larva approaches hungrily from the other. As if that were not enough for the scale, a parasitoid fly lays its egg on another victim just behind.

their parents to look for more scales. Soon the human operators of the factory scoop up the wasps (which are completely harmless to people) and seal them in plastic vials about the size of your index finger. Hundreds of these vials are then taken into the field. In orange orchards in southern Queensland, for example, each bush has about a dozen vials attached to scattered branches. An experienced biological control manager can attach a vial and whip out the stopper in one deft move, so distribution of the wasps is rapid. The orchards in this region are free of pesticides in most seasons.

One theme in this chapter has been that the services provided to us by friendly species most often occur out of sight, or at best on a microscopic scale. This is especially true of the nematode worms, which are mostly very small, transparent cylinders with almost no distinguishing features. Yet around the world nematodes are raised to control pests in situations as varied as slugs in gardens in England and larvae that kill trees in China and Australia. Bananas are highly vulnerable to herbivorous weevils that are attracted to natural wounds in the banana stem. Imaginative scientists are experimenting with injection into the wounds of a fluid containing live nematodes, so that when the weevils arrive they are immediately infected with deadly control agents. The story is often gory. For example, the flies that infest pigs can be attracted to traps where nematodes lurk, waiting to climb a leg of the fly as it feeds, entering via the anus and gradually consuming the fly's internal organs. The battles are fierce—but on a Lilliputian scale.

Tinier still are the viruses, bacteria, and fungi used to combat pests. These microorganisms form the fourth great *P* group, the pathogens, after the predators, parasites, and parasitoids. It is likely that all species on Earth are vulnerable to one pathogen or another, and the race is on to discover viruses, bacteria, or fungi that can be put to work attacking pests. Many viruses known as baculoviruses naturally attack insects. The scientists who study them (virologists or molecular biologists) are exploring this ultraminute world for biological control agents. Safety is a special concern, as no one wants a rogue virus that will run wild over beneficial as well as destructive insects. Many fungi naturally attack insects, and major research efforts are under way worldwide to harness them to control beetles that attack crops such as sugarcane, potatoes, and peanuts; domestic pests such as termites; and agricultural pests such as plague grasshoppers.

One of the great advantages of fungi (and bacteria) is that the control agents are minute and can therefore be sprayed onto a crop or onto the soil, where pests can be reached once the bacteria or fungi spread. Again, on a microscopic scale the battle can be dramatic. Take as an example a fungal spore that lands on the back of a pest beetle. Beetles are well known for the tough material that constitutes their exoskeleton, so the fungus has difficulty getting inside. However, over millions of years of evolution, a wild solution has emerged among the natural insect-attacking fungi: the spore germinates and grows into a short tube ending in a small pad equipped to apply immense pressure. Research has estimated that if the equivalent pressure were available to a human hand, it could lift a school bus with ease. Once the fungus has penetrated the beetle's armor, the tube enters the beetle and hundreds of branches develop to penetrate its internal organs.

Even the rhinoceros beetle, which grows up to 10 centimeters long, sports a huge horn on its head, and is a major pest of coconut crops, is susceptible to fungal spores that can be sprayed on the plantation. Fungi can be used against weeds and even against other fungi. Antifungal fungi can be utilized in the natural form, or the chemicals they secrete to attack other fungi can be synthesized and applied to crops such as apples, grapes, and sugar beets.

Weed-attacking fungi are at work in the United States, Canada, Israel, India, Africa, and Australia against enemies with names like sicklepod, jointvetch, cocklebur, milkweed vine, and Russian thistle. Finding the right fungus is not easy. The pathogen must perform exactly when required in the field, in all kinds of weathers; it must be simple and cheap to grow in the factory, and easy to package or spray. Many promising species have had to be abandoned because they failed one or more of these tests. Yet the industry still has explored only the tip of the iceberg that represents the fungal diversity of the world.

It is impossible to *overestimate* the value of natural enemies. We simply cannot get along without them. In their absence, virtually all tropical crops would likely be destroyed by herbivores as human attempts at chemical control were overwhelmed by the evolution of resistance to the pesticides. Growing crops in temperate zones would be much more difficult despite the natural control of insects that winter cold supplies. At the very least, world food production would plummet at a time when the need to increase that production is critical. The upshot would be massive famines, social disruption, possibly even large-scale warfare. Whether civilization could survive a loss, or even a great diminution, in the scale of the services it receives from natural enemies is an open question. Those "enemies" continuously supply one of the least celebrated wild solutions.

EIGHT
Naturally Selected

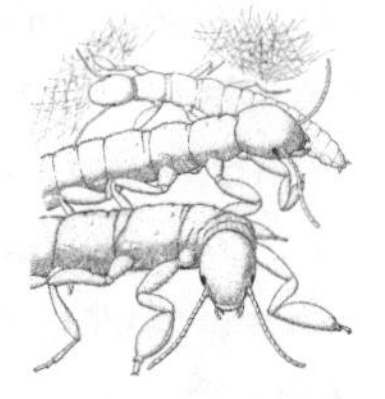

The Australian bull ant is one of the largest and fiercest ants in the world. As you approach a nest, the guards become agitated, probing the air with their antennae, following your movements with dark, bulging eyes. The alarm spreads and workers pour out of their holes, falling over each other in their eagerness to attack. If one reaches you, its huge pincer jaws dig deep into your skin and its abdomen curls around to bring the sting into play. Jaws and sting together create an excruciatingly painful wound and you would do well to retreat before the rest of the colony joins in.

What enemy could such an animal possibly fear? One potential answer is the echidna, a small Australian mammal that traps ants, termites, and other insects on its long sticky tongue. As it digs out an ant nest, it is protected from the attacks of the inhabitants by a dense coat of hair and spines. However, anyone who has watched an echidna knows that it does not have everything its own way. The ants fight back as the nest boils over with enraged workers and soldiers. Although the echidna mops up many of them and may dig for stragglers, it soon shuffles off, long snout twitching, beady eyes sparkling, leaving a rear guard of ants to repair the damage and, as in many ant and termite species, take the dead to the colony graveyard.

While the echidna is one of the great enemies of ants, it rarely threatens a colony with extermination. Colonies are damaged, but they recover. The real enemy is far more dangerous and can wipe out a nest completely. The enemy that most threatens the ants is not immediately obvious, yet it is the same enemy that most threatens humans, an enemy we all fear—disease. What we have in common with the bull ant, and with all ants, is that most of us live together in large numbers. The ants have their colonies and we have our cities and towns. Whether ant or human, the grave danger when we live close together in large numbers is contagious disease.

Now, ants have been around many millions of years longer than humans. How have they coped with disease? Their solution, although ancient, seems stunningly modern: they have evolved glands that secrete antibiotics. There are two glands per ant, each nestled just above a back leg. Inside, where the leg muscles enter the body cavity, is a chamber where a milky fluid accumulates. The chamber leads to the outside by a short tunnel through which the fluid moves and oozes over the surface of the ant. The entire structure is known as the metapleural gland, and its contents, chemicals known as metapleurins, are of great interest.

Until recently, the purpose of these glands and the fluid they produce was a mystery, but groups of scientists in Germany, the United States, and Australia have gradually uncovered the story. The secretion can be extracted by means of a very fine glass pipette and tested for antibiotic activity (Figure 23). A simple experiment reveals its potency. In one test tube a culture of the fungus *Candida* is mixed with metapleural secretion, while the fungal culture in a second test tube is left untouched. To both test tubes we add a fluorescent dye that is rapidly absorbed by the fungal cells. After a couple of hours both cultures are placed under an ultraviolet light to stimulate fluorescence. One tube, the one with no added secretion, glows bright green; the cells, still growing and vigorous, produce an enzyme that activates the dye. In contrast, the second tube never glows because the secretion has killed the fungi and the cells cannot produce enzymes to fire up the dye.

This test, and many others, have shown that the fluid is a power-

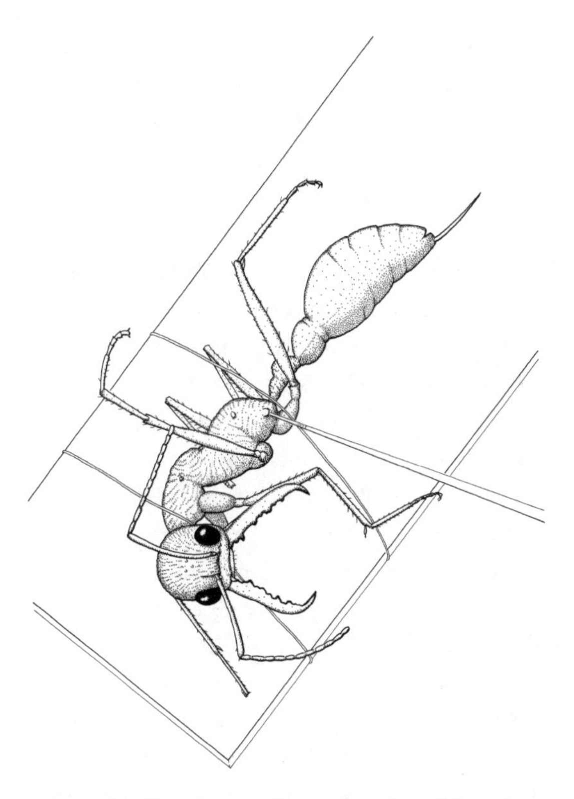

Fig. 23 A bull ant is secured to a microscope slide so that a fine pipette can remove secretions from its antibiotic gland. The treatment is harmless, and the ants are returned to their colonies.

ful antibiotic. One striking result of the secretion is that the surface of the ant is almost free of bacteria and fungi; indeed, it is much cleaner than most human skin.

A strange aspect of this secretion is that although the queen and her workers have metapleural glands, males do not. In fact, if males are kept isolated from the ant workers in the laboratory, they soon become badly infected. Of course, this experimental situation is unlikely to occur in nature, but we can inquire why the males have no metapleural glands. The answer may be that males are relatively short-lived and so have little use for the glands. The young males appear only with the onset of the breeding season, maturing just as the virgin queens take flight. Having mated, the males are excluded from the nest and so become immediately vulnerable to the crowds of predators attracted by the swarms of mating ants. While in the nest, the males may obtain antibiotic secretions from the workers that tend and feed them, but once on their own and having mated, they fall prey to predators or disease microorganisms.

It seems that many millions of years ago, when the first primitive ant colonies evolved, it was advantageous to produce a chemical defense against disease—a strong antiseptic to keep marauding bacteria and fungi at bay. Contagious disease, by definition, thrives where animals get together. In fact, it would seem that the production of some kind of antibiotic would be absolutely necessary to the development of any kind of society. That the glands in ants are so large and produce such copious amounts of antibiotic strongly suggests that microbes were the ants' major enemy and that natural selection would have quickly eliminated colonies without this protection.

The antibiotic gland is an adaptation to disease, and disease is a problem ants and humans share. To be sure, the actual bacteria and fungi that attack us are different from those that attack ants, but the challenge is the same: to kill them before they kill us. We now ask what may seem a very strange question: inasmuch as ants and humans share this problem, is the adaptation that ants have evolved of any use to us?

If we think about it for a moment, it is very likely that the ants are on to something. There have been many millions of generations of ants. This is an important point, because the production of a new generation means the production of new variation. This is evident in the different faces of the grandparents, parents, and children in a given family. Each generation has a new set of variations: a slightly broader nose, curlier hair, a new curve in an eyebrow. In exactly the same way, there must have been generation upon generation of variation in the chemicals produced by the antibiotic glands in ants. In the process known as natural selection, these would have been tested against the disease each colony encountered. Chemicals that proved less effective would have gradually disappeared as those ants failed to combat disease and died. By contrast, those colonies that produced the most effective antibiotic variants would have lived to reproduce and pass the genes for those antibiotics to their offspring. In this way, countless variations on the gland and its antibiotic contents would have been tried over hundreds of thousands of generations. The best would have survived.

All of this rather obscure biology becomes much more familiar if we put it in modern production terms. The research and development time has been approximately sixty million years, millions of prototypes have been tested, and the entire research and development program was free—courtesy of evolution by natural selection. This is a modern industrialist's dream come true. The question now is, How much of this dream can be put to use by human beings?

Understanding this situation, some alert biologists have explored human applications of the antibiotics that ants produce. One metapleural antibiotic has been patented as an antiseptic and is used in hospitals. The focus on ants has yielded another patent resulting from the surprising discovery that ants not only defend themselves by means of antibiotics, but they also have an immune system. That fact had not been anticipated in the textbooks. An Australian student studied the blood of bull ants as they became infected by invading bacteria. By comparing the "fingerprints" of the proteins before and after infection, he saw clear differences as mysterious proteins appeared in the blood once the ant was diseased. He knew that in the human immune system, the next step would be the appearance of new chemicals manufactured to combat the bacteria—antibodies. Sure enough, the unfamiliar ant proteins not only had strong antibacterial activity but were new to science and worth patenting. A bonus was that chemical analysis identified which piece of the protein was responsible for killing the bacteria. When the student chemically removed this piece, the molecule did nothing; but when he put the whole molecule back together, the antibiotic effect was restored.

The filing of these two patents prompts a rather startling conclusion: ants are a valuable resource! They certainly do not fit the usual image of a resource such as oil or iron ore. But if oil and iron ore are mineral resources, is it not reasonable to think of ants as biological resources? If oil and iron are manifestations of mineral wealth, could ants not be manifestations of biological wealth? The fact is that exploration of antibiotics from ants has so far proved commercially viable. While the idea that ants are important commercially may seem absurd, their usefulness has been established beyond doubt by the challenging legal and business protocols required to establish international patents. It is evident that ants, with their head start of sixty million years, know a lot about combating disease that we humans do not. With this in mind, it will be interesting to delve further into the concept of antibiotics as wild solutions.

Before we do so, let us pursue a small diversion. A few years ago the story of the bull ant antibiotic was told on Australian television. The next day, one of the scientists involved received a phone call from an old aboriginal woman. She recalled her childhood in the Australian outback and the times when she, or her brothers and sisters, were cut or their skin grazed. Her mother would take a clean cloth and toss it on a nest of bull ants, stirring them up with a stick. When the cloth was covered with ants, the stick was used to hold it aloft and shake it until no more angry insects were attached. Then the wound was bound with the cloth, in the sure knowledge that no infection would occur. This conversation reminded the scientists of the treasury of knowledge held in the minds of indigenous people, like a living library. It also struck them that their discovery of the bull ant antibiotic had been made centuries before, by another human culture with a history longer than theirs by as much as forty thousand years.

To return to bull ants, they are not the only social insects; in fact, all ant species are social (except for a few that are parasites) and there are probably at least another 9,500 species in the world. What different potions have they concocted over evolutionary time to fight disease? The leaf-cutting ants that inhabit the rain forests of Central and South America cultivate a fungus for food. Now this may seem odd, but the extraordinary thing is that their underground fungus cultures are both warm and moist, ideal places for scores of different fungi to flourish. Yet their fungus crop is a monoculture, relatively uncontaminated by other fungal species. It is the equivalent of a garden that never has weeds. The metapleural secretions are mildly antibiotic, but the ants carry bacteria on their bodies that secrete antibiotics specifically adapted to the nest environment. Relatives of these bacteria, the Streptomyces, are the ones that produce many commercially important antibiotics.

The possibilities are even greater. Ants are only one of four major groups of social insects; the others are bees, wasps, and termites. Antibiotics have been found in bees and termites, some far stronger than any yet isolated from ants. The fungus-culturing ants that inhabit warm and tropical regions of North, Central, and South America are replaced by fungus-culturing termites in Africa and Asia. No one knows how they keep their fungal cultures pure, but antibiotics very likely play a part.

There are still other kinds of social animals, such as certain spiders and beetles, but we do not know how natural selection has provided them with defenses against disease. It will be worth finding out, for the pressing reason that in the everlasting battle between humans and microbes, it is the microbes that are winning. We humans have been manufacturing antibiotics for about fifty years and they have provided such powerful defense against microbial disease that we have taken them for granted. But all along, the disease organisms have been evolving their own defenses against our antibiotics and now are poised to overcome them. Already antibiotic-resistant tuberculosis has regained a hold on portions of the human population, and other bacteria are proving extremely difficult to treat. The bacterium known as hospital staph, or golden staph, for example, remains a serious problem in hospitals. Many patients who recover from surgery suffer lingering and resistant golden staph infections.

You may well ask, Why don't antibiotic-resistant microbes attack ants? The answer is that they do. Figure 24 shows a common disease of ants, a fungus (called *Cordyceps*) attacks the internal organs of its victim. As the disease gets worse, a remarkable thing happens. The fungus affects the brain of the ant, altering its behavior in such a way that when it is about to die, it climbs a plant and, in its death throes, bites the stem hard with its jaws. It dies in that position. Soon afterward the spores of the fungus emerge from the corpse, borne still higher by a long stalk. It seems unlikely to be an accident that the spores, which are dispersed by the wind, find themselves conveniently up among the air currents rather than below ground.

Ants do have diseases, but co-evolution is an arms race. Sometimes the fungus or bacterium is ahead, at other times the ants are winning. Today *Cordyceps* is a serious problem for some kinds of ants, although they do not appear to suffer from many other diseases. We do not know why this is so, but it will be fascinating to find out—if for no other reason than it may help us. One possibility is that ants make a variety of antibiotics and that resistance to one component of an antibiotic cocktail is quickly overcome by another antibiotic. We know that the bull ant has two major kinds of defense against microbes: antibiotics and an immune system. Thus it seems reasonable to expect chemical diversity in both.

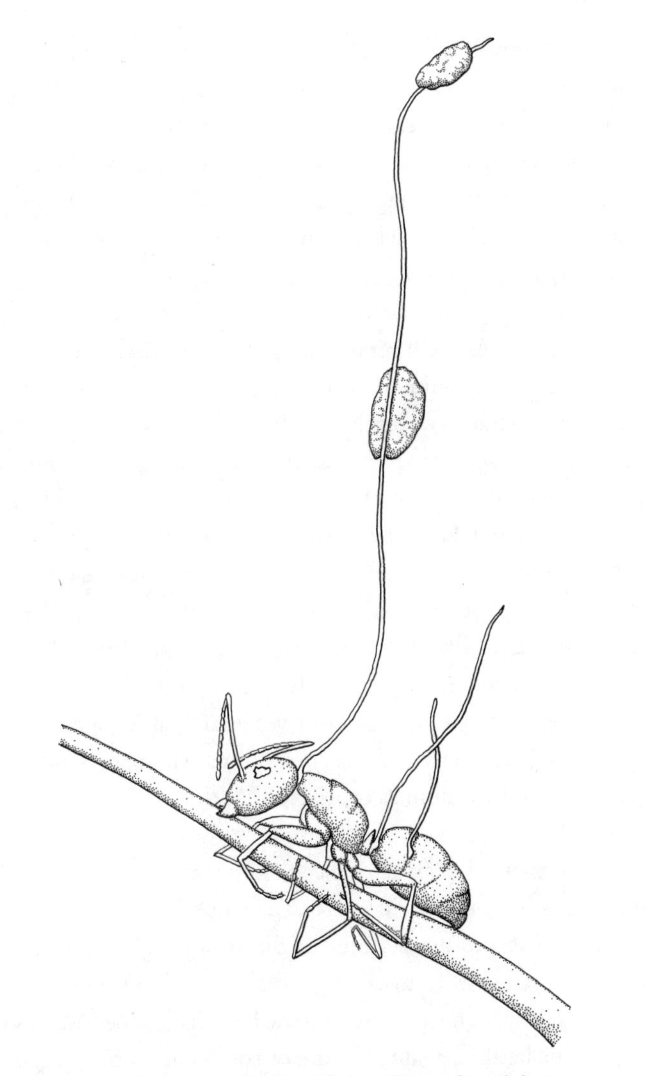

Fig. 24 Ants too suffer from disease. Here a fungal disease has invaded the body of the ant, gradually digesting its internal organs. The stalk carries the fungal spores into air currents.

Drug companies are looking for innovative sources of antibiotics for clinical use, and the same problems are developing in the veterinary world. Beyond the care of human and animal patients, new antibiotics are needed to protect crop plants. Beyond those is a vast array of industrial products ranging from wool and cotton to fuels and paints, all of which are attacked by microbes and must be defended against them.

At the heart of the quest is a powerful question: *Where have antibiotics evolved?* As we have just seen, one answer is that we expect them among other animals that, like us, inhabit complex societies with many thousands or millions of individuals living together. Natural selection has made animals such as ants, bees, and termites potentially useful to us in a totally unexpected way. If we think carefully, we may be able to take advantage of their adaptations to disease.

If we pursue this question further, more unexpected answers come to mind. For example, what about animals that live in dirty places? But first, what do we mean by the word "dirty"? Often we mean locations where there is already a very active microbial community such as corpses, dung, or rotting vegetation; although such places may be repugnant to us, they are food or home to many kinds of animals. How do these animals cope? How do they deal with the microorganisms?

Think about the larvae of flies that live in dung or corpses—in other words, maggots. They may experience two threats from microbes: the threat of disease and the threat of competition. After all, the microbes are present to eat the dung or corpse too, and frequently it is a race between the maggots and the bacteria for the best pickings. Whether the battle is against disease or competition, the maggots are likely to have an armory of antibiotics.

Evidence for antibiotics in maggots has existed for many years, as it was observed long ago that those that eat rotting flesh tend to keep battlefield wounds clean. In the days before efficient medical care, solders lying in the field with open wounds were often found to be in relatively good condition. The maggots ate the dead tissues and the raw surfaces and created alkaline conditions that inhibited bacterial growth. Initially no antibiotics could be found, but perhaps no one looked in the right place. The bactericides were discovered in the maggot's wastes, adding another macabre twist to the story. While to most, the sight of the wriggling masses is repulsive, the evidence for antimicrobial activity is undisputed. Even in more recent wars, when medical services were disrupted, soldiers have continued to be saved by maggots. Today, when wounds in modern, high-tech hospitals are invaded by deadly antibiotic-resistant bacteria and fail to heal, surgeons still successfully use maggot therapy as a last resort.

Maggots have been the source of some antibiotic patents, but these have been derived from a handful of common, well-studied species. There are many hundreds of thousands of other species in the world (some with the unlikely name "sarcophagus flies") and antibacterial agents have been discovered in their larvae. Some bees defy the popular image of the healthy life, that of visiting flowers and collecting pollen for their children back in the hive. The so-called filth bees collect dung and carry it back to their nests. It seems reasonable to assume that these bees protect their larvae from the seething microbes already inhabiting and decomposing the "groceries." The same kind of protection is to be expected among the hundreds of species of beetles that specialize in burying corpses so that their larvae may live in them.

Without belaboring the point, we have not yet considered the many thousands of different kinds of animals that inhabit rotting vegetation such as leaf litter or the compost heap in your garden. These include earthworms, roundworms, flatworms, millipedes, centipedes, the larvae of flies, larval and adult beetles, crickets, springtails, a myriad of single-celled animals and plants, bacteria, fungi and slime molds—to name just a few. It is also worth considering seeds in soil and litter. How do they protect themselves? The intriguing thing is that some don't. In fact, some need the bacteria and fungi to erode and decompose the hard seed-coat to begin the slow but carefully timed process of germination. Others, however, must last for years before the appropriate conditions for germination appear. Many seeds that fall to the forest floor find themselves where little sunlight penetrates and where most of the water and fertilizer is monopolized by the biggest trees. No use germinating in these conditions, so the seeds wait to spring into action until a tree falls through storm damage or old age. They may wait for decades or even centuries before the chance occurs, in an environment full of microbes that must be kept at bay. What has natural selection provided in this situation?

The question of where antibiotics have evolved has still more possible answers. Many natural products appear to offer a "free lunch." Nectar is the most obvious example; it is full of nutrition, especially sugars. Still, the nectar is not free for the pollinator, such as a bee or hummingbird or honey possum, because it must work the pollination system of the flower to obtain its reward. Bacteria and fungi are not beholden to the system, and if they can reach the nectar, they will colonize the nutritious substance and consume it. How many plant species protect their nectar with antibiotics? Recognize that decomposed nectar is likely to be repellent to a hard-working pollinator, so the premium on keeping nectar fresh is high. So far we know of only a few kinds of nectar that contain antimicrobial chemicals, but there are about two hundred thousand more plant species to investigate.

Nectar is, of course, a plant product and plants have been a significant source of antibiotics in cooking for thousands of years. These antibiotics are usually known as spices which, when added to food, not only make it more tasty, but slow down the rate at which the food goes bad. A greater variety of spices are added, especially to meat dishes, in hotter and more humid regions of the world. Yet the strongest antibiotics do not necessarily come from the tropics, as garlic, onion, oregano, and thyme all appear at the top of the list. It is not obvious why some plant species contain more, or stronger, antibiotics than others. Perhaps they live in particularly "dirty" locations, that is, places where microbial activity is very high. Plants that exist for much of their lives as bulbs in composted soils may be examples. Alternatively, the chemicals that we call spices may be directed at repelling herbivores such as caterpillars or deer, and the antibiotic properties may be merely a useful sideline. Many carnivores, including dogs and cats, will now and again chew a particular plant species. This behavior, which often puzzles pet owners, may be an attempt to self-medicate against bacteria or perhaps parasites. By following wild animals, especially those that are sick, field biologists have begun to draw up lists of plant species that may have healing properties of pharmaceutical interest.

To return to the theme of free lunches, it should be emphasized that nectar is not the only free lunch in nature; there is another that is extremely common. In fact, if you look around the room right now, you may see it: spider silk. It is a complex substance, mostly protein. The possibility that the spider may defend its web against microbes was hinted at by a Reverend Topsell in England, who wrote in the year 1607 that a skin injury bound with spider's web "binds, dries, glutinates and will let no putrefaction continue long there."

Spider silk (Figure 25) has many remarkable properties, as we shall see in a later chapter, and one of these may be the inclusion of antimicrobial chemicals. There are several hundred thousand species of spiders in the world that capture their prey in an almost unbelievable variety of ways. There are hunters that outrun their victims; others that leap, trailing a silk line like a bungee jumper in case they miss and fall; spiders that wait in a silk tunnel to ambush their prey, either by throwing open a trapdoor or by stabbing through the tunnel wall with their long, curved fangs; those that run on the surface of ponds

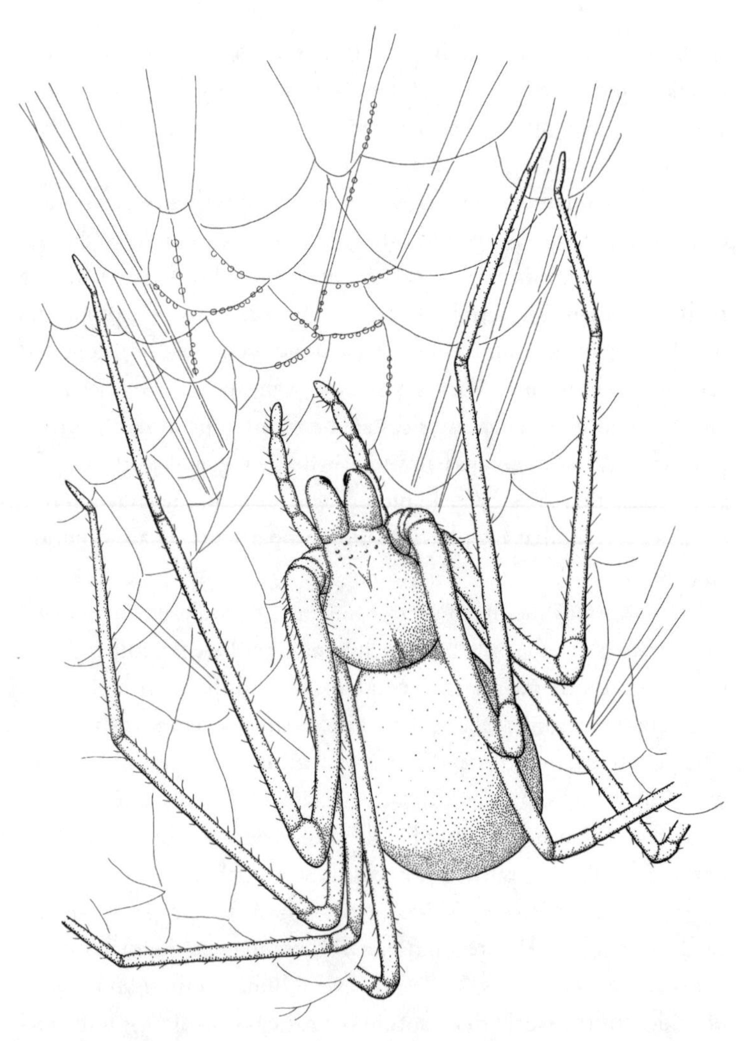

Fig. 25 Antibiotics from a cave spider? A cave spider hangs in its microbe-resistant web. The moist walls of its cave are festooned with fungal growth, but the silk it uses to make its web and egg cases is strongly antifungal.

and lakes; some that cast a silk net over their prey as though fishing; and still others that whirl a blob of glue over their heads on the end of a strand of silk.

Amazing though these hunting methods may be, our present question is, how long is the silk exposed to microorganisms and under what conditions? Two illustrations will help. A sensible recycling device used by many spider species is eating the silk spun the night before if it has not been used, that is, if it has failed to catch anything. In this case we might anticipate that the silk contains few, if any, defenses against microbes, even if the spinner lives in warm, moist tropical forest. By contrast, many ambushers construct silk tunnels in the ground that not only last for years, but nestle in soils swarming with bacteria and fungi, even in cool climates. It is reasonable to suggest that this silk is well defended against microbes.

Aside from the huge variety of spiders, many other kinds of animals spin silk—and for many different purposes. Pseudoscorpions, or false scorpions, are so called because they look like the real thing except that they have no tail and no stinger. Rarely more than the size of a fingernail, these dainty relatives of spiders glide among the leaf litter, apparently without effort and in any direction, looking for all the world like miniature ballet dancers with eight legs, holding their diminutive claws above their heads. The claws are for grabbing minute prey, but they also produce silk to make dense, protective sacks for the eggs. Next on the list of silk spinners are caterpillars. The silkmoth is familiar, but many thousands more moth and butterfly species spin to make structures as varied as cocoons to protect pupae, harnesses to hold pupae in position, and communal tents within which hordes of caterpillars feed on leaves.

Next we come to a group of spinners unknown to most people, including biologists! These are the foot-spinners, or in scientific language the Embioptera (Figure 26). These small insects live exclusively in silken tunnels in leaf litter, under bark, or in rock crevices. Their

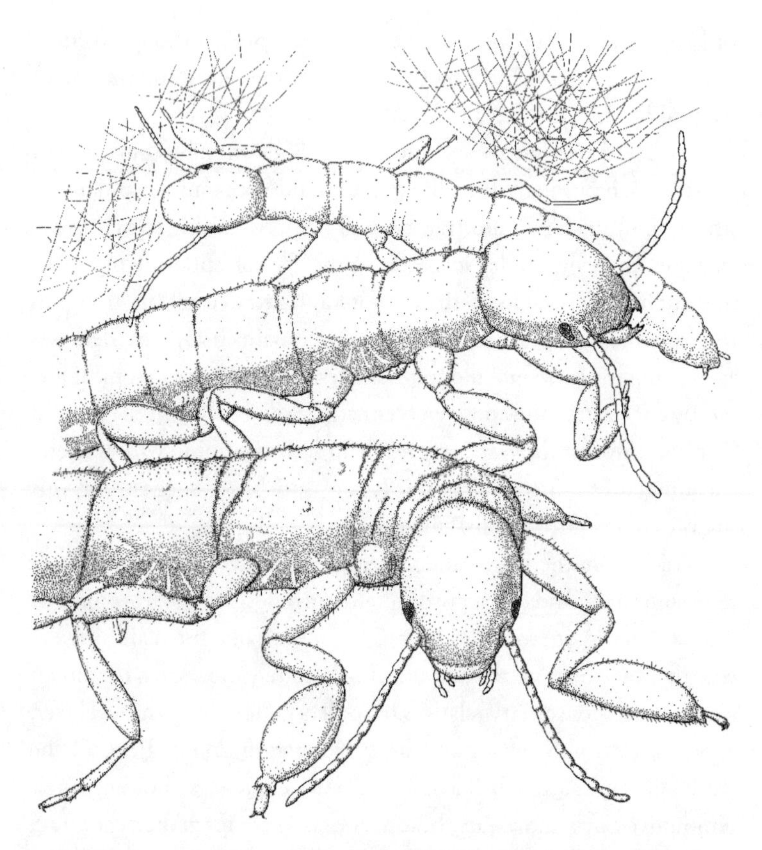

Fig. 26 A group of foot-spinners, their silk-producing glands located in the swollen parts of their front legs. These animals live under logs and in moist litter where many microbes thrive. How do they defend their silken tunnels against decay?

front legs are the reason for their name. The end joints are greatly enlarged to house a mass of silk-producing glands, so that the neat termitelike animal appears to brandish a pair of boxing gloves. Many individuals live together in extensive webs with numerous tunnels. Are these silken homes protected by antibiotics? Nobody knows. Only about two hundred species are currently described, but biologists working with these animals suspect that there may be as many as two thousand species scattered throughout the hot and warm regions of the world.

A final line of thought on the origin of new antibiotics is the protection of delicate surfaces. Many animals lay eggs without shells, often in masses containing hundreds or thousands of these highly nutritious morsels. Studies show that snail eggs, which in many species have no shells, contain antimicrobial substances in the outer layers. Even more intriguing is the discovery that some female centipedes coat their eggs with a fungicidal secretion, and queen fire ants spray their eggs with venom that has strong antibiotic properties.

A completely different set of delicate surfaces are those used for breathing. A variety of water beetles and water bugs secrete antimicrobial substances that keep their precious breathing mechanisms free of bacteria and fungi. Who would think to look into the breathing tubes of a water bug for antibiotics? This example shows yet again how diverse nature is and what a variety of potential sources of useful substances there may be, if we only have the imagination and knowledge to find them.

One of the major high-tech advances in pharmaceutical research has been the use of powerful computers to construct hundreds of thousands of chemicals in the hope that some will emerge as new antibiotics. There is no doubt that chemists have put this technique to good use and have come a long way toward designing specific molecules targeted at specific microbes. However, it is interesting to ponder whether nature, through millions of years of trial and error by natural selection, has not already sieved through all the possibilities and identified the best of the lot. We wonder if the antibiotics found in nature—in ants, termites, bees and wasps, beetles and caterpillars, nectar and silk—might not be an appropriate place to start. Then the computers might take over to tinker, refine, and develop the raw materials.

The history of pharmaceutical exploration of natural products reveals many failures, perhaps because the searchers have not asked suitable questions to help focus their research. Maybe the questions posed by ecologists and evolutionary biologists, and the way they think, will turn the tide and increase the success rate. In some kinds of mining, it is considered normal for only one of every thousand test shafts to "come good." Conceivably, with the clever application of biological knowledge to pharmaceutical problems, that rate will be surpassed.

NINE

Miners' Canaries and Sentinel Pigs

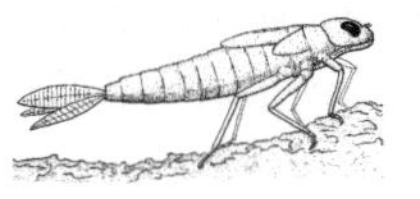

Pollutants and wastes from industry and agriculture as well as from our homes are now worldwide and many countries have passed laws to stem the tide. While the passage of such laws is an appropriate step, it presents a major challenge: how to monitor the environment to make sure the laws are effective. What we call the environment is both vast and complicated and includes a wealth of different habitats: soils, forests, deserts, rivers, lakes, coastal waters, deep oceans, and the atmosphere. In other words, it is the entire surface of the planet, whether liquid, solid, or gaseous. How can we monitor on such a vast scale?

At this point we should introduce the miners' canary. Mining below ground is very dangerous, not least because of the presence of extremely poisonous or explosive gases, many of them with no noticeable smell. In the days before the invention of modern instruments to detect these gases, miners took caged canaries into the tunnels to act as early-warning systems. Canaries sing loudly and often, their silence was cause for worry. If it went on too long and the birds appeared to be distressed, there was a strong possibility that poisonous gases were about, and it was time to leave.

The idea that a living organism can monitor the environment is not new. However, the concept is being developed in ways never previously imagined. Many different species have been shown to be potential monitors and thousands more are waiting in the wings. In this chapter we demonstrate that the enormous challenge of monitoring can be met by selecting organisms that act as miners' canaries for each different environment. The beauty of this approach is that most environments harbor an immense variety of potential monitors; thus, monitor species can perhaps be tailored to specific problems. Which monitors we select depend on many factors, as we shall see, but we start by agreeing that we will not use species that are rare or endangered.

In river environments scientists have successfully found a variety of monitoring species (see Figure 27). One of the best known is the mussel, a humble mollusk with a tough shell, hinged down one side so that the other side opens, like a door, to expose the feeding organs to the river water. This arrangement is similar to the shell of the scallop (which may be more familiar because it is a more popular seafood). When alarmed, the mussel snaps shut via a powerful muscle. It is this muscle, together with some of the internal organs, that creates the tasty mouthful that many people enjoy. However, it is not the delicious taste that concerns us here, but those feeding organs over which the river water passes. The feeding mechanism filters food particles from the water and these are transferred to the animal's stomach, thereby providing the monitoring system.

The food particles are a mixture of bacteria and single-celled or very small plants and animals. These organisms have themselves been feeding in the river, taking in pollutants with the water. The pollutants have lodged in their minute bodies, sometimes killing them. Alive or dead, they are swept into the mussel's filtering mechanism, which acts like a conveyor belt and delivers them into the digestive tract. There powerful enzymes break down their bodies into individual molecules, and these are the mussel's meal. The very same process also releases the pollutants, which are then absorbed into the mussel's body tissues as though they were food. The result is a mussel laden with several kinds of pollutants.

To put it another way, the mussel is a monitoring device, for it has gathered a sample of polluted river water and stored the pollutants. This is a process called bioaccumulation, a word that says it all: *bio* = organisms, *accumulation* = gather up. Thus, pollutants in the water

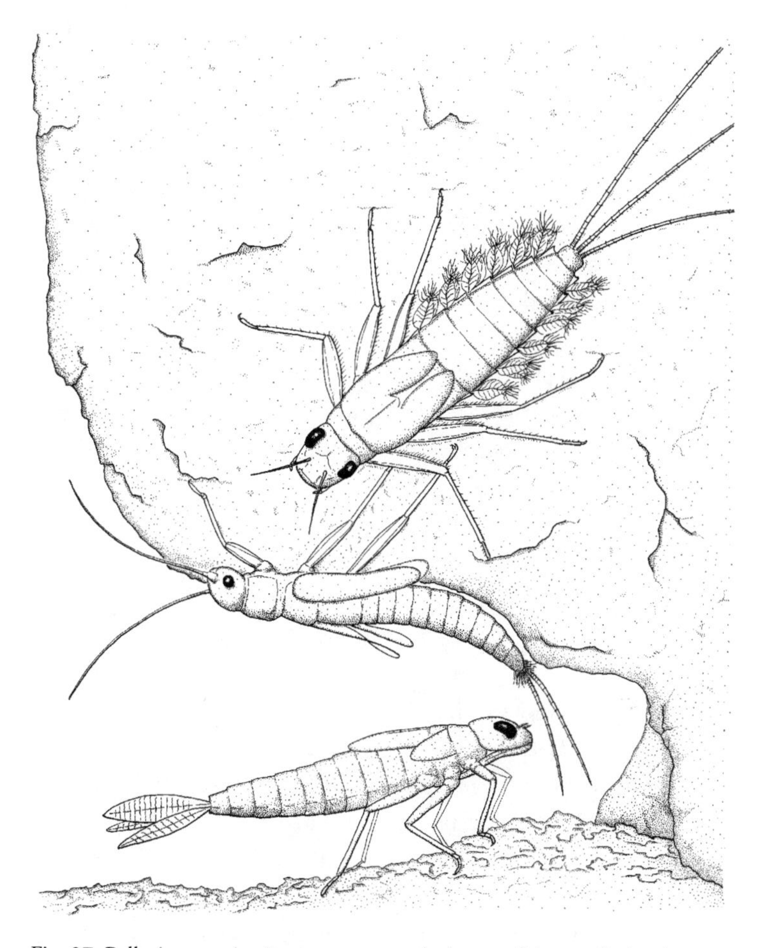

Fig. 27 Pollution monitoring instruments: the larvae of the mayfly (top), stone fly (middle), and dragonfly live in streams where they absorb pollutants. Some common species can be collected and taken to the laboratory to determine levels of pollution in their bodies. When these levels are mapped, it is often possible to pinpoint the source of the pollution.

have been gathered up by the mussel's food and, in turn, gathered up by the mussel as it feeds. The original dose of pollution in each tiny food item is relatively insignificant, but a cluster of them in the mussel's stomach delivers a much larger dose—hence the concept of accumulation. You may believe that it would be smarter to sample animals higher up the food chain, such as the water birds that eat the mussels. Surely they would have a much larger bioaccumulation in their bodies. Indeed they do. But birds are relatively less common than mussels, so that if they were frequently collected their numbers would soon decline. (In any case, it is doubtful that the public would allow the collection of water birds for the laboratory.) To return to the mussels: once they have been collected, bringing them into the laboratory for analysis is relatively straightforward.

This is the core of the biological monitoring system. It has everything we need. First, the mussel cannot avoid sampling the river water because it has to feed. What is more, it feeds most of the time, so it is sampling most of the time. Second, mussels tend to be very fertile species and where they occur, they are often present in large numbers. So if they are collected in reasonable quantities, there will always be plenty left. Third, they occur over large areas, which means that they can be collected from carefully chosen locations. Thus, a series of mussels taken downstream from a riverside factory may show low levels of pollutants far away from it but high levels close by. These levels can be mapped to determine the source of the pollutants.

Sometimes the presence of pollutants in the animal's body can be detected only by extremely sensitive laboratory instruments—as when the animals have been collected far from the source of the pollution, or when the pollutants are widespread but highly diluted. At other times the effects of pollutants may be obvious because the animal is dead or deformed. Pollutants from a factory or sewage outfall cause ripple effects ranging from badly damaged animals close by to subtly poisoned ones far away. Recent developments have resulted in the ability to biopsy or sample the gills without harming the animal.

Biological monitors such as mussels form the basis of new industries that monitor specific environments for a wide variety of pollutants. However, the benefits of these monitors are not limited to their biological properties. Imagine trying to locate the source of a pollutant in a complex river system such as a delta, or in a series of interconnected lakes. If sophisticated instruments needed to be installed at regular intervals, the expense of designing and manufacturing, installation and maintenance, would be enormous. By contrast, the mussels have been designed, manufactured, installed, and maintained by nature. They monitor in order to live. The main cost is for collecting and analyzing them. As long as the collections are made wisely so that the mussel populations continue to reproduce and thrive, the biological monitoring system remains active.

The science of biological monitoring is most fully developed for aquatic environments such as rivers, streams, and lakes—and the sea, especially in coastal environments. One reason is that the animals can be brought into the laboratory in order to develop exquisitely sensitive tests. Wild mussels relocated to an artificial stream in the lab can be exposed to specific pollutants and then examined to discover precisely what happens. Frequently the pollutant is stored in the animal's tissues and can be measured directly. Sometimes, however, a chemical causes malformed larvae, organs, or appendages. A deformed biomonitor may be the victim of any number of pollutants from a variety of sources. However, lab tests can show that a particular malformation arises from exposure to a specific pollutant, thereby raising the possibility that the source can be found. Just as there was once a close connection between a silent canary and poisonous gases, mussels and other animals can flag a problem with similar accuracy.

Numerous species have been considered for biological monitoring in aquatic environments. In fact, it is now more or less obligatory to monitor several species simultaneously. These are chosen to represent different sorts of organisms and life styles. A group of freshwater biomonitors will often include a fish, an amphibian such as a tadpole, one or two invertebrates that float or swim in the water (these may include the tiny crustacea known as water fleas, or midge-fly larvae) and a couple of invertebrates that crawl about on the bottom such as the larvae of damselflies, mayflies, stone flies, or dragonflies. These four "flies" are not really flies at all, but have been given the name because their aquatic larvae metamorphose into winged adults that hover, dive, and swoop above bodies of freshwater. Usually the larvae are used as biomonitors because they absorb pollutants through their feathery gills or with their food. However, the presence of a rich variety of adults in the air above a stream or lake is often interpreted as a sign of clean water.

Muddy habitats contain other animals that process mud to live and that are therefore potential biological monitors. These include a wide variety of small worms with the tricky name oligochaetes. The name means "not many bristles," and these worms differ from their relatives, the polychaetes, which have so many bristles they are often known as bristle worms. Many oligochaetes inhabit the mud at the bottom of ponds, lakes, and rivers, where they are freshwater equivalents of earthworms. They spend their lives feeding on the bacteria and other minute organisms that live in the mud. Pollutants, absorbed into the worms' bodies, are detected when the animals are collected and taken to the lab for testing.

Before we proceed to look at biological monitors for ocean water, we should backtrack a little and elaborate on some organisms we have just mentioned: bacteria. We usually think of them as agents of disease or as food for larger organisms, but do they hold any promise for monitoring? They do, and many people are aware that their water supplies are monitored for sewage contamination by looking for coliform bacteria, a polite term for the bacteria associated with the dung of humans and domesticated animals. A high coliform count usually indicates severe problems in the water supply, and residents are advised to boil their water to kill the bacteria. Meanwhile, the source of contamination may be associated with a coliform "hot spot," a stretch of river or a region of a reservoir with an elevated count of the indicator microbes. A variety of bacteria are being explored as biosensors. Species that can fluoresce either naturally because they provide the glow on luminous deep-sea fish or artificially because they have been genetically engineered to contain a gene (for example, from a jellyfish) that causes fluorescence. Several laboratories in the United States and Great Britain are exploring the possibility of using bacteria (or yeasts, which are single-celled fungi) to glow in the presence of specific pollutants. Possible applications include the detection of heavy metals in the ocean, chemical weapons in the air, and carcinogens in food.

Bacteria can also be used to evaluate the quality of soil. The best-known examples are those that regulate the principal natural fertilizer in the soil, nitrogen. These bacteria absorb nitrogen from the air, where it is abundant, and transfer it to the soil, where it is often in short supply. This astonishing process has been going on for millions of years, with these bacteria one of the mainstays of agriculture. Unfortunately, the addition of artificial fertilizers can disrupt the process, making it difficult to determine which natural, native nitrogen-grabbing bacteria should be present.

Among the better-known candidates for biological monitoring in the oceans are the polychaete worms mentioned a moment ago. All over the world these animals are collected from the sands and muds of beaches, bays, and estuaries and used as fish bait. Because they are also of major scientific interest, several species have been brought into the laboratory to study their responses to pollutants such as pesticides, heavy metals, and wastes from oil spills. (We often marvel at the use of the word "spill" in this context. For the most part, a spill is a small event that occurs, for example, when a glass of milk is accidentally overturned, yet the same word is used to describe the tidal waves of deadly black ooze that ravage entire coastlines for decades!) Polychaetes from seawater can be analyzed for pollutants in much the same way as mussels from freshwater.

Other possible biomonitors include shrimps, prawns, crabs, sea urchins, starfish, clams, fish, and bioluminescent bacteria. Sea squirts are being investigated intensively as they are common seashore animals, generally clinging to rocks near the low-tide mark. They look like clumps of stumpy, leathery fingers, each of which contains a large filtering mechanism for the removal of food particles from the water. At low tide, when some animals may be exposed, they still contain seawater. It is their habit of squirting that water if you touch them that gives them their name. Sea squirts can be cultivated in the laboratory and their responses to common ocean pollutants such as oil studied in detail. Thus, when an animal is brought in for analysis, the state of its environment can be determined from the condition of its organs. A very different type of monitoring has been carried out with seabirds: the contamination of the oceans by mercury over recent decades has been traced by analyzing the rising levels in the birds' feathers.

Biological monitoring of soil is not as far advanced as that of water. Since many pollutants dissolve in water, they diffuse extensively in aquatic systems. The monitoring species literally inhabit a vast chemical bath, and exposure to pollutants is more or less inevitable. By contrast, the nature of soil changes greatly from one place to another, so that it encompasses many different kinds of environments: pockets of leaf litter, boggy places, sunny and shady areas, stones, logs, bodies of dead animals, and patches where soil is thin and close to the bedrock. An effective soil monitor must somehow summarize all these differences.

Earthworms are promising candidates, as they eat soil for food, digesting the tiny organisms that live in it. While soil, or dirt, may not seem very nutritious, it is usually packed with bacteria, fungi, and minute animals and plants, jammed together with tiny fragments of compost and minerals. We know that earthworms absorb pollutants along with their food; in locations contaminated with copper, lead, zinc, and cadmium, their body tissues are loaded with these heavy-metal pollutants. As with mussels, collection of earthworms from carefully mapped sites can help pinpoint the source of pollutants.

Another candidate for monitoring soils is a land-living relative of the shrimp, the humble sow bug, also known as the pill bug or slater in various parts of the world. Pill bugs are usually about 2 centimeters long and are very common in the surface layers of the soil, among dead leaves and twigs and under rocks and logs. They are useful partly because they are so common, especially in urban areas, and partly because they absorb pollutants such as heavy metals. The two features together mean that they can be collected from roadsides, the median strips of freeways, and industrial sites to monitor levels of pollutants such as lead that are emitted from cars, trucks, and chimney stacks.

Are there species that monitor the atmosphere or the quality of the air? The answer is yes, of course. Most of us have seen plants turned from green to yellow or brown because of atmospheric pollution in big cities or industrial areas. In these situations it is often very obvious that the air is polluted. There is little need to examine the leaves on trees when the smell of smog, itching eyes, or wheezy breathing are excellent indicators of air quality! Of greater interest are organisms that, like the miners' canary, reveal danger when it is invisible—organisms such as lichens, honey bees, and yeasts.

Lichens are familiar to most people as gray-green crusts on rocks, tree trunks, and roof tiles, or as pale green beards that festoon twigs and branches. They are often thought of as "lower plants," regarded as somehow inferior because they do not have flowers or never grow very large. In actuality there is nothing "lower" about them, and they are not really plants. Lichens are unique, physically intimate partnerships between fungi and algae. Under the microscope the crust or beard is revealed as a fungal body, consisting of mats of intertwined fungal hairs called hyphae, among which minute, single-celled algae or cyanobacteria nestle. While the organism that the partnership produces, the lichen, is both small and slow growing, it withstands some very harsh conditions, particularly drought. Lichens can be baked all summer long on the exposed surfaces of rocks or up on your roof and appear dry, brittle, and dead. Yet after rain, when they absorb moisture, they spring back to life.

Curiously, the toughness that gets lichens through the driest of times frequently does not help them survive air pollution. This characteristic is the basis for lichen biological monitoring. Different lichen species are vulnerable to different kinds of air pollution, especially when it is dissolved in the water they absorb at the end of a dry period. Some are extremely sensitive to the common pollutant sulfur dioxide and are never found close to industrial smokestacks. In fact, these and related lichen species can tell us much about the air we breathe. One study revealed a lichen "desert" close to an industrial complex, with the number of lichens increasing with distance from the complex. It works in reverse, too. In cities with effective regulation of air pollution, lichens sensitive to sulfur dioxide are migrating downtown.

There are many other ways that lichens inform us about the atmosphere, some species being vulnerable to acid rain and others to airborne heavy metals. Research in northeast Italy suggests that the relationship between air quality and the diversity of lichen species is sufficiently close that counting the number of different kinds of lichens that survive in your district can indicate your chance of contracting lung cancer. Elsewhere, lichens that absorb radioactive elements can be used as a record of environmental disaster following accidents at nuclear power plants or atmospheric testing of nuclear bombs.

The science of lichen biomonitoring is well established and has encouraged the search for similar kinds of environmental sentinels in plant groups. Mosses are serious candidates as they too are very common and are sensitive to specific atmospheric pollutants. As with lichens, some mosses bioaccumulate toxins and can be analyzed in the laboratory to reveal patterns of pollution from power plants, mines, and smelters.

Mosses, like lichens, are immobile and therefore record air quality at a very local level. Is there a biomonitor that moves around, sampling the air as it goes? The answer is yes, and the monitor is the honey bee. A forager collects pollen and nectar that may be contaminated, but also the hairs on its body electrostatically attract dust particles. Thus, a bee may collect or attract pollutants as it carries out its routine chores. Also, worker bees may fan out over several square kilometers in search of food, but all of them return to the hive, where they can be collected and analyzed. Research has shown that bees taking nectar from plants growing on contaminated soils reveal high levels of arsenic and cadmium in their bodies. This fact is particularly useful when the plants are crops. The body hair on honey bees collects dust, which around nuclear power plants and weapons factories may be contaminated with radionuclides such as cesium 137 and cobalt 60. The bees, normally associated with the production of a favorite food, honey, are being investigated as biological indicators for leaks of radioactive materials or toxins from waste dumps.

The third set of biological monitors of air quality belong to a group of organisms already encountered in this chapter: the fungi. Within this vast kingdom of hundreds of thousands of species, the group known as the yeasts is most often thought of in the context of making bread or brewing beer. But yeasts are widespread, occurring naturally in diverse places such as the bark of trees, the nectar of flowers, and the surface of leaves. In fact, the surface of leaves is home to a great variety of yeasts and the monitoring aspect arises from the fact that some, like lichens, are extremely sensitive to atmospheric pollutants such as sulfur dioxide. Monitoring is a matter of taking swabs from leaf surfaces, growing them in laboratory cultures, counting the yeast colonies that appear, and mapping the results. This process is sometimes helped by the fact that many yeast colonies are bright pink.

Most examples we have discussed so far involve the use of a single species as an indicator or monitor. However, in many cases groups or communities of species may perform this function. The advantage is that the responses of several species can yield a more complete picture of environmental disturbance. This is especially true when the species are involved in several aspects of the ecosystem being monitored, for example, eating many different kinds of food, having many different sorts of homes, and foraging in many different parts of the habitat. Animal communities such as these also provide the opportunity to detect and monitor effects other than pollution, such as logging in forests and tourist presence on coral reefs.

In many countries, mining companies are now required to restore the land after mining has ceased. Few believe that the ecosystem that was removed will ever return, but at least a functioning ecosystem, presumably with some of the native flora and fauna, will come back to cover the slag and tailings. Some mining companies take this duty very seriously and not only reuse saved topsoil but plant it with native herbs, shrubs, and trees. In most cases, not enough time has elapsed to know whether the infant plant community will persist. The plants may be surviving, but for how long? Is the soil community of bacteria, fungi, and invertebrates that generates their nutrients returning? Are the animals that pollinate their flowers and disperse their seeds reappearing? Have the food chains that support these animals been restored?

These are complicated questions that involve hundreds, even thousands, of different species of all shapes and sizes, and it is not possible to monitor them all. Scientists have searched for biological monitors that somehow summarize all aspects of the system by virtue of being *involved* with all of them. Ants are prime candidates because they are diverse, abundant, and deeply involved with many of the other organisms that live in the same location. For example, many species nest in the soil, others in hollows of plants. Some forage on the ground, others in the tops of the trees. Some ambush invertebrate prey in the soil and leaf litter, others actively hunt in vegetation. Some sip nectar from flowers or honeydew secretions from greenflies, others eat seeds or gather leaves. In other words, ants are part of virtually everything that is going on. Ant mimics add fascinating evidence for the long-term, intimate associations between ants and other animals. Why? With ants often dominant in terms of both variety and numbers, other animals including spiders, bugs, beetles, and mantids mimic their bodies and behavior in order to blend into the landscape and escape their attention. Ants as biological monitors depend on the notion that with the ant community involved in so many aspects of the ecosystem, changes in that community will reflect changes in the ecosystem.

This concept may prove effective. In mine sites undergoing restoration, a gradual buildup of ant species over the years is usually evident. At first just a few species, the ant equivalents of weeds, move in as soon as possible. These are slowly augmented and replaced. Over time the ant species more characteristic of mature, stable habitats appear, and eventually the rarer species. In many cases these last return only when their specialized prey or nesting places have been restored. Although this progression is not surprising, the stages in the restoration can be carefully studied. For example, the appearance of particular groups of ants can be correlated with particular events such as the establishment of shrubs, tree saplings, or leaf litter on the ground. Study of these events repeatedly and at different sites makes it possible to associate particular groups of ants with particular environmental conditions.

Former mines are not the only kinds of disturbances that have been researched, and ants are not the only candidates for biological monitoring on land. The effects of fire, air pollution, grazing, and logging have all been monitored by ants, and detailed pictures are beginning to emerge for certain environments. It is now possible to lay out a series of ant traps in an apparently uniform area of grassland or forest and, by analyzing the catch, identify plots that were once logged, grazed, or burned. And ants are not the only indicators. Spiders, scarab beetles, tiger beetles, springtails, and fungi are all being considered, either as backups to the ants or as monitors in their own right.

It seems that butterflies and birds have especially strong potential. First, they are relatively easy to identify. Second, they are much loved, popular animals that most people relate to. Third, because of this affection, they are studied by all sorts of people; large organizations of amateurs and professionals have as their principal goal to observe the animals and record when they are doing well and when they are not. There is an important fourth reason why butterflies and birds are valuable biomonitors: like ants, they sample the environment thoroughly. As an example, observation of thirty butterfly species flying in a patch of forest over the course of a week may immediately suggest that the plant community is in good shape. Most of the caterpillars of those thirty species eat different plant species and each of those species will have to be present to produce the caterpillar's butterfly. Birds also sample the plants, sometimes requiring specific flowers, fruits, or seeds. Or they may specialize on plant structures such as tree hollows that form only in certain tree species. Many reflect the structure of the vegetation itself, preferring particular habitats such as treetops, inside the foliage near the trunk, tall shrubs, low shrubs, or the grasses and herbs close to the ground. Thus, a variety of birds reflects many positive aspects of the environment. To put it another way, a high diversity of bird species strongly indicates the presence of a high diversity of other species in the ecosystem that supports them.

Researchers now consider it wise not to rely on a single group of monitors, but to analyze a variety of species. Butterflies, moths, birds, and ants have high potential not only because they are abundant and diverse, but also because they reflect different aspects of their food chains. Continuous monitoring also helps avoid misleading conclusions. For example, in remnants of rain forest left after logging or burning, the variety of moths and butterflies can remain high for some time. Sometimes, though, the truth is that the food plants of the caterpillars gradually die out for reasons completely unrelated to the caterpillars. A single sample of insects may present a picture of ecosystem health, whereas a series of samples over several years will reveal a gradual decline.

In some circumstances it is not necessary to track the biomonitors species by species. In freshwater communities it is possible simply to know what the animals do for a living—what their life style is like—to determine what is going on. Invertebrates that inhabit lakes and rivers can be classified as leaf-shredders, leaf and rock scrapers, water filterers, predators, or general scavengers. A bad case of pollution may eliminate the water filterers that rely on relatively clean water and increase the proportion of scrapers feeding on elevated levels of algae clinging to submerged rocks and logs. In these circumstances, scientists may avoid having to identify all the species by simply classifying the animals according to life style and drawing conclusions from the changes therein. Like changes in the cars in the driveways when neighborhood property values rise, so too life styles shift when pollutants arrive.

An animal group proving useful for monitoring is yet another kind of worm, the nematode. These worms are extremely common but often overlooked, not just because most of them are tiny, but also because most of them are, frankly, rather ordinary with no bristles or any other distinguishing external features. Yet in river muds and ocean sediments, the species that make up the nematode community vary according to the level of pollution. In pristine river habitats one set of species occurs; where the water becomes polluted, some species disappear, or one or two species that formerly were rare become common, even dominant. In some highly polluted rivers, only one or two nematode species survive. In extreme cases, a single species takes over, becoming so overcrowded that worms unable to find room in the mud form pulsating balls that roll along in the current.

All the information in this chapter tells us the requirements for an effective biological monitor. It should live throughout the area to be studied. It should be easy to collect and analyze and sufficiently common that collecting does not harm its populations. Its responses to specific pollutants should be understood and reliable. It should not move around too much, so that its responses can be pegged to a particular locality. It is helpful to be able to tell the age of specimens, as missing age groups provide useful information. It helps too if the species is economically important, or is related to a species that is, because that makes it easier to obtain funding to get the job done properly! A desirable group of species for biological monitoring (ants, nematodes, butterflies) should be widespread; it should also contain a great variety of species with highly diverse life styles or resources, and the relationships of the group to pollutants and other disturbances should be well understood.

Early in this chapter we mentioned that malformations in biological monitors can reveal pollution. Unfortunately, some pollutants may be so widespread and so apt to cause malformations that a biomonitor is scarcely needed to reveal it—almost any animal will do. This is the case with the broad range of chemicals released into the environment that mimic hormones. The best known are the "gender benders" that are so close to sex hormones that they play havoc with development and reproduction. These chemicals may be present in paper, paint, plastics, pesticides, and other materials, and therefore in thousands of manufactured products. Manufacturing processes and waste-disposal systems leak them into the environment, where they contaminate the food chains of the world and affect animals as diverse as shellfish, fish, whales, gulls, and humans. Common malformations in animals include complete gender changes and many variations thereof, including animals with both male and female genitalia. In human beings, the evidence is compelling that these pollutants lead to a wide variety of conditions and diseases. Theo Colborn and her team have reported the situation in their book *Our Stolen Future.*

In this chapter we have discussed the various equivalents of miners' canaries. What about the other half of our title, the sentinel pigs? These are monitors for the detection of diseases that affect humans and domestic animals. In the case of Australia, they are sentinels because they are located on offshore islands and provide long-distance warning. There are a hundred islands in the Torres Straits, the narrow sea between the northern tip of Australia and Papua New Guinea. About fifteen of them are inhabited on a regular basis and the Torres Strait Islanders, in collaboration with the Australian government, keep pigs as monitors. Flattering or not, pig physiology is sufficiently like human physiology that by keeping track of their diseases, we can keep track of our own. When pigs are exposed to a mosquito-borne disease such as encephalitis, their immune systems manufacture antibodies in the blood to fight it. A small blood sample is taken from the pigs every month and checked for these antibodies. Their presence indicates disease, and by mapping the results it is possible to detect southward movement of mosquitoes toward the mainland.

You may wonder about the health of the Torres Strait Islanders; if they live on the same islands as the pigs, there is not much early warning for them. Quite true. As Australian citizens, they are immunized against the diseases. Why then is immunization not done routinely on the mainland? The answer is because the cost is too great. There are 19 million mainland Australians and many millions more domestic animals. Further, for some diseases the injections have to be given more than once. In fact, the turnover of domestic animals as they are sent to market means that immunization has to be repeated with every generation. The monumental cost, in terms of both money and labor, makes it cheaper and more efficient to maintain sentinel animals.

With the early warning provided by the pig's immune system, most medical and veterinary emergencies are confined to a relatively small area.

One final detail concerns the living conditions of the pigs. Their pens are located under palm trees that rustle in the balmy sea breezes. The pigs snooze or root about in the sand and coconut husks, and several times a day a villager passes by and tosses in leftovers from the family meal or some other delicacy. Not a bad life!

TEN

Chemical Engineers

Just for a moment, imagine that you are a plant. Now imagine that a moose is about to take a mouthful of your youngest leaves—or a large, bristly beetle grub is about to bore into your seedpod to dine on your offspring. Running away is impossible because you are rooted in the ground. So how do you defend yourself? Prickles, thorns, and tough bark are one way. Chemicals are another. If you can manufacture poisons in your tissues or, even better, chemicals that repel attackers before they take a bite out of you, then you may be able to avoid serious damage.

Back in the world of people, we know that spices and herbs are chemical mixtures in plants that we use to flavor our food. They include cinnamon, pepper, mustard, nutmeg, oregano, cumin, dill, fennel, and mint. Other chemicals derived from plants are medicines, used to treat diseases as varied as cancer and malaria. In fact, the plant world is a vast drugstore that is being explored by many of the world's largest pharmaceutical companies. Some natural chemicals such as the ingredients of curries and chili sauces span the culinary and medical worlds, as they both stimulate our taste buds and kill contaminating bacteria. But many harmful plant chemicals such as nicotine and caffeine and poisons such as belladonna or deadly nightshade cause every conceivable type of effect on the human body—vomiting, diarrhea, blindness, headache, and heart attack. Chemicals from plants are among the favorite poisons of crime and mystery writers because they are so varied and often so hard to detect in the corpse!

The knowledge that plants contain both useful and harmful chemicals is ancient, but not until recently have people started to ask why the chemicals are present at all. In order to find out, biologists have performed a great many experiments to discover what the chemicals do. In one type of experiment, varieties of a plant species that differed according to the level of a particular chemical in their leaves were exposed to the animals that normally ate them. Most of the time the animals were caterpillars or slugs, easily raised with their plant food in greenhouses. The experiments were revealing: in most cases, the varieties with highest levels of the chemical in question were not eaten as much as the varieties containing smaller amounts.

It was not long before it was realized that these observations and experiments provided a window into an astonishing chemical world war that has been raging for millions of years. Most of the chemicals we enjoy as spices, herbs, flavors, and medicines evolved as toxins to repel plant enemies. While we find them useful, herbivores find them repugnant, distasteful, or poisonous. This is not hard to imagine if you have ever had a mouthful of succulent food with too much pepper or mustard. It probably brought tears to your eyes, led to a fit of coughing, and nearly made you choke. Even nutmeg, which to most people is a relatively gentle flavor, in large doses becomes a poison. We are comparatively big animals and usually can recover quickly, but for smaller animals, a dose of one of these chemicals seriously discourages any further contact with the plant it came from.

We know that plants make chemicals of many kinds, and that many of them either prevent herbivores from grazing them or poison them if they do. Just how adept can plants become? Some have crude but extremely effective poisons such as cyanide gas that are released as soon as they are bitten. Others wage a far more subtle war. Their leaves contain chemicals called tannins which work in a particularly sneaky way. Imagine a chocolate bar on a hot day, totally melted into the wrapping paper. Now convert that wrapping paper into thick cardboard. You have a mess that is 10 percent chocolate and 90 percent cardboard, but the only way you can get at the chocolate is to eat the whole thing, cardboard and all. That is the tannin strategy. The tasty and useful parts of the plant are surrounded by tannins with almost no nutritious value at all, but the herbivore has to eat the whole business to get any food. As this means that most of the time the herbivore's stomach is full of useless chemicals, you can recognize that the animal's growth is stunted, along with the chance of producing a lot of offspring.

Some plants undermine their insect herbivores by manufacturing animal hormones. Unlikely as this may seem, a well-known example is the insect molting hormone that regulates the growth of insect larvae. The plant concocts this crucial hormone in its leaves, and although the larvae eat the leaves, every mouthful contains a chemical that retards the ability to grow up. This fiendish strategy means that after the larvae have fed happily for a few days and outgrown their old skins, they attempt to molt. However, their bulging bodies are confused by the powerful hormone taken in with their food and the old skin will not come off. The result is fatal. True, the plant has lost some leaves, but next year the ranks of invading herbivores will be much thinner.

This process may seem a lot of effort just to fight off a mere insect, when it seems that the real threat is surely the "big guys" such as deer, giraffes, elephants, buffalo, or the vast herds of antelope that roam the African grasslands. While these large animals are of course menacing, they are in fact relatively insignificant compared to the real hordes of herbivores such as caterpillars, leafhoppers, blackfly, shield bugs, nematode worms, termites, leaf-cutting ants, slugs, and snails, to name just a few. These animals make up for their small size with their sheer variety and numbers. For example, there may be twenty species of termites in a block of tropical grassland, comprising billions of individuals that weigh more than all the big herbivores put together. And these armies of mostly small herbivores have evolved a myriad of ways to attack plants. They munch roots, chew buds, suck twigs, bore stems, slice leaves, steal nectar, ransack flowers, and devour seeds. They have been the enemies of plants for millions of years, so over almost unimaginable lengths of time, plants have had to become chemical engineers to repel, deceive, and poison their foes.

These examples quickly reveal that chemicals govern the relationships between plants and their enemies. However, the study does not end here. Hundreds of thousands more natural chemicals have evolved as weapons or as the means for survival in harsh environments, and still more attract mates, catch prey, or fight disease. In this chapter we look at some of these chemicals and consider how they may provide wild solutions.

Before we leave the battleground between plants and their enemies, let us take a brief look at some natural plant chemical defenses that have a promising future. Many plant chemicals are already used as pesticides: nicotine, pyrethrum, and rotenoids. Thousands more are known and, beyond those, thousands more await discovery. One example is the neem tree from Burma. Someone noticed that neem seeds seem to be attacked less than the seeds of other trees, and this observation prompted an investigation to find out if they contained any natural pesticides. It turned out that neem seeds were full of them. The chemists gave them names that sounded like alien invaders: azadirachtins, meliantriols, salannins, and numbidins. Field trials in which seed extracts were sprayed on crops in liquid or powder form showed that, either way, they kept insect pests under control. In fact, they seemed so effective against the pests but so harmless to birds and mammals that pesticide companies in several parts of the world raced to isolate and patent the relevant chemicals. As a spin-off, a wide variety of pharmaceuticals were also discovered, leading one scientist to speculate that the neem may become the most valuable tree crop in the world. Other plant species that seem to escape attack include the green ash tree, the tiny herb known as wintergreen, and many members of the mint family (including catnip). These are just two of many more plant species that are being investigated for commercial pesticides.

Plant extracts that kill pests are only one weapon in this complex chemical battleground. Researchers are exploring some very complicated trickery to defeat crop pests. In one experiment a crop pest was placed in one branch of a Y-shaped glass tube and nothing in the other. A predator of the pest was allowed to walk up the stem of the Y, so that it had the choice of a meal or no meal. The branches of the tube were plugged with cotton so that the predator could not see what they contained, and air was gently wafted through them so that any scent in the branches of the Y could be detected by the predator. In almost all cases, the predator went straight toward the branch containing the pest, suggesting that the pest produced a smell attractive to the predator. In further experiments, when intact leaves of the crop plant were placed in one branch of a Y-shaped tube and pest-damaged leaves were placed in the other, the predators almost invariably went toward the damaged leaves. In a final twist, it was found that the droppings of the pest were sufficient to attract the predators!

It turns out that many plant species, including crops such as cabbages, corn, and cotton, release odors from the wounds inflicted by herbivores as they munch leaves and stems. These scents are the plant equivalent of alarm bells or distress signals. From the point of view of the predators and parasites, it makes sense that these odors are the very ones that attract them to the pest, but it has taken us humans a while to realize that scents attractive to predators and parasites of herbivores might make effective pesticides. In other words, we may be able to protect crop plants by taking advantage of the enemies of *their* enemies. Research is isolating and purifying the chemicals that produce the attracting odors. Who would have thought that the smells of wounded plants, plant pests, and the droppings of plant pests might lead to important commercial agricultural products?

In a slightly different scenario, many herbivores use the smell of their food plant to find it. It is not hard to imagine, from a herbivore's point of view, that if the plant it desires is just one of a hundred different species growing in a meadow or forest, it is particularly useful to be able to sniff it out. However, this behavior may be a weakness that can be exploited. For example, aphids that feed on beans are attracted by bean smell but are repelled by cabbage smell because they are unable to feed on cabbage. Then why not spray a field of beans with cabbage smell, and a field of cabbages with bean smell? Agricultural scientists are currently assembling cocktails of chemicals that they hope will both repel pests and attract their enemies.

The science of pest control is becoming even more devious. Imagine a caterpillar eating a plant, going about its daily life, doing what must come naturally such as quietly depositing its dung on a nearby leaf. A necessary, normal activity, but one that is a matter of life or death for the simple reason that odors released from the droppings are the very perfumes that attract the caterpillar's enemies, especially wasps (Figure 28). This is the situation for at least two major crop pests, the corn earworm and the cotton boll weevil. It is possible that commercial products that save billions of dollars worth of crops will be developed from the chemicals in caterpillar and weevil droppings.

All of these examples are based on the knowledge that insect pests have evolved in a complex and subtle chemical environment, and that biological chemists thus have opportunities to combat them in ways that are more environmentally friendly than old-fashioned pesticides. Traps containing synthetic female-attracting chemicals lure males to a sticky end, the smell of food may be sprayed on items the pest cannot eat, and the natural defense odors of their food plants may be boosted to attract swarms of enemies.

So far we have talked only about the chemicals used by plants to defend themselves. What are the possibilities for animals? Many people have heard of army ants, which travel in columns of millions on

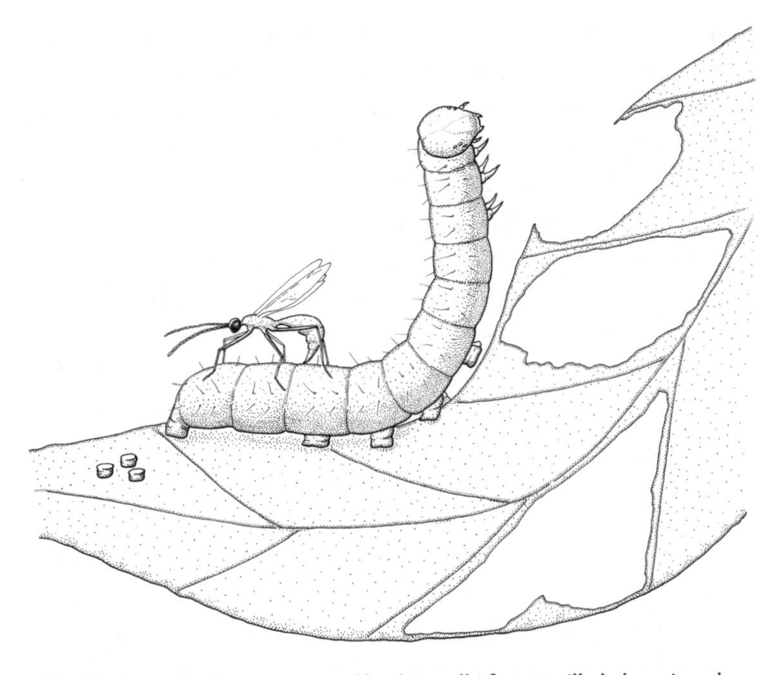

Fig. 28 A parasitoid wasp, attracted by the smell of a caterpillar's droppings, lays an egg on its victim. Many crops are protected from pests by this mechanism.

the floor of tropical forests. We experienced an invasion of these animals in Trinidad and learned how awesome but useful they are. Awesome because the animals on the forest floor scatter as the column of swarming workers and soldiers advances. Those that cannot escape are trapped, stung to death, and carried off as food. Useful, because as they approach your house, you carefully wrap your food in ant-proof boxes and simply let the ants proceed. A couple of hours later, when they are gone, you find your house swept clean of pests from the smallest tick to the largest tarantula.

The hunting strategies of social insects such as ants and termites are not limited to brutish, overwhelming numbers. In fact, they have a massive array of armaments, many of them chemical. Among the chemical weapons are those found in Monomorium ants. Often very small and apparently not aggressive, they seem to hold their own among larger ant species and among their ancient enemies, the termites. Detailed research on these tiny animals has shown that they secrete repellent chemicals that must be immensely potent. Only minute amounts are produced, but they have devastating effects such as making larger ants run away and killing termites. One such chemical has been analyzed with a view to commercial production as a new-generation termiticide.

The possible sources of chemicals that repel pests emerge only through a detailed knowledge of natural history. For example, wasp nests are very familiar, especially those that dangle from a branch or rafter on a slim stalk. But why should the connection between the nest and the structure from which it hangs be so thin? Wasp nests vary greatly in size, from that of a ping-pong ball to that of a watermelon, but the stalk is rarely thicker than that of an apple. Biologists interested in the social insects have learned that the most implacable enemies of social insects are other social insects! We saw just now that ants (social insects) have defensive chemicals that are extremely effective against other kinds of ants and termites (more social insects). And ants are among the greatest enemies of social wasps (yellow jackets and their kin). Careful observation of wasp nests has shown that ants would very much like to attack, but access is limited to that slender stalk. Moreover, as the wasps make the stalk they repeatedly smear it with a secretion from their backsides. Sure enough, that secretion is a powerful ant repellent.

In Chapter 8 the leaf-cutter ant was introduced as a possible source of antibiotics, but its chemical world may be even more sophisticated. The fungus that the ant cultivates in underground "gardens" grows on the leaves collected by the ants, thus transforming plant material that the ants cannot digest to fungal material that they can. In an astonishing twist, it appears that the fungus releases chemicals that tell the ants that the leaves they are collecting are unpalatable to the fungus. Having received the message, the ants collect different leaves that with luck are more to the fungus' taste.

We know that leaf-cutter ants are major pests of crops and plantations, destroying hundreds of millions of dollars worth of production every year. Question: Where might a leaf-cutter ant repellent evolve? Answer: In its own fastidious fungal gardens. Might this chemical be used to spray crops, telling the ants that those plants should not be harvested?

The opportunities for harnessing natural chemicals for human use appear to be endless. Although we cannot discuss them all here, we should mention a few more. The first is the natural armaments of predators, especially animals such as spiders and scorpions. Advances in molecular biology now enable researchers to obtain from these animals not only the poisons but in some cases also the genes, the genetic blueprints that govern manufacture of the poisons. This means that chemists can work on understanding the structure of these poisons and how they may be altered, if necessary, before being deployed among crop plants. Of course, many will be too complicated, others perhaps too lethal, but if we remember that there are more than one hundred fifty thousand species of spiders and scorpions, the choice is not exactly limited. And we have not begun to consider other predators such as ants, termites, wasps, centipedes, and a huge variety of beetles.

Chemicals that repel insects rather than killing them are more desirable in a variety of situations, especially when killer chemicals—pesticides—wipe out useful insects or contaminate the environment. Insect repellents are made in nature by some beautiful insects whose wings, when at rest and folded over their backs, make them look as if they are wearing lace. These are the lace bugs. Observations suggested that although they appear to be a tasty meal for birds, as many insects are, lace bugs are scrupulously avoided. Careful experiments showed that they secrete chemicals that birds intensely dislike. Biologists believe that these insects, and other kinds that birds avoid, might be the source of repellents suitable for protecting crops, especially fruit. They argue forcibly that such repellents would not harm birds, which is a great advantage to wildlife management in the tricky situation where valuable food crops are attacked by much-loved or even endangered bird species.

The pupal stage in insects is especially vulnerable because the mobility of the larval stage is lost, and the wings of the adult stage are not yet present. Some pupae are anchored with silk to a twig or rock or some other object while the animal inside undergoes the magical transformation from larva to adult. Pupae are extremely vulnerable to attack and so, rather like plants, which also cannot move away from enemies, some defend themselves with chemicals. Research into a species of ladybird beetle has shown that its pupae manufacture insect repellents. What is remarkable about this tiny animal is that the repellents are composed of an array of starting molecules or building blocks, and these are combined in various ways to yield a wide variety of end products. The diversity of repellents made in this way presumably deters a variety of enemies and may also slow the evolution of antidotes. It is as though ladybird beetles have invented the modern science of combinatorial chemistry, in which chemists use computers to build complex molecules by changing the ways the building blocks are put together. The research is still in its early stages, but beetles may well represent a huge library; the species are the books, and the information they contain is the key to a wealth of successful chemical invention.

A completely different kind of chemistry is involved with glue. What is the natural history of glue, and how might it guide us to better adhesives? It has been known for many years that velvet worms and some spiders catch their prey by squirting instant glue at them. The action is so fast and the glue so instantaneous that a fly or mosquito is snared before it can escape. This is remarkable. The spitting spider does not sneak up on a fly, take out a tube with a nozzle, and ask its prey to hold still for a few seconds until the glue dries. Instead, at one moment the glue is fluid in the animal's body and in the next microsecond it has traveled through the air and entangled the fly in a solid coil, sticking leg to wing, antenna to foot, and head to ground. One species of spitting spider specializes in capturing jumping spiders. This may not sound very dramatic until you recall that jumping spiders are themselves fierce predators adapted to pouncing on their prey at extremely high speeds.

The velvet worm is armed with what can only be described as a pair of glue guns, one on either side of the mouth (Figure 29). The jets of glue that emerge have been filmed and the instantaneous nature of the glue has been confirmed by many biologists. However, when Christine Turnbull was drawing the velvet worm you see in the figure, she was not prepared for the animal's next trick, when some of the glue went astray and coils of it appeared around the worm's own antennae and forelegs. If you have ever used fast-setting glues, you are well aware what a nuisance it is to get some on your fingers. More often than not they stick together and are separated only with great difficulty. Objects that become enmeshed are often stuck so fast that they break before you can pull them apart. Our velvet worm, however, contorted its front end to pass its antennae and forelegs through its mouth; when that was done, the glue was gone. Thus the velvet worm has not only evolved an amazing instant adhesive, but it has also developed a harmless solvent that works with equal speed.

Adhesives such as those of the spitting spider (Figure 30) and the velvet worm have at least one commercially important feature: they set very quickly. Other glues in nature have even more spectacular features. Have you ever wondered how oysters, barnacles, and tube worms stick to rocks even though battered by the waves on the seashore? Whatever the glue, it must work on wet surfaces, be resistant to the cor-

rosive effects of seawater, and be able to withstand the pounding of endless waves. Both the tiny larvae and the adult animal with its heavy shell have to cling for their very lives.

One trick the larvae use is to swim toward an area of rock already inhabited by marine bacteria. They can smell them and swim to the place where the odor is strongest. The bacteria themselves must be able to stick to their home, and they secrete a very effective adhesive to keep from being washed away. When the larvae first attempt to settle, they use the bacterial secretion to glue themselves to the rock, but as they mature they produce their own. This adhesive has been collected and analyzed for potential commercial use, especially for underwater repairs such as on ships' hulls and the submerged parts of oil rigs. Future uses may include repairing damaged organs such as the liver that have soft, moist surfaces not easily stitched.

In other cases the larvae make their own adhesive. The protein adhesive that mussels secrete is known to be immensely strong and able to penetrate and stick to all kinds of surfaces, including wood, metal, teeth, bone, and even teflon. Research continues on the bioadhesives produced by the larvae and adults of barnacles and many other mollusks and tube-living worms of seacoasts and even the deepest parts of the oceans. In fact, as much as twenty-five hundred meters below the surface, volcanic openings known as hydrothermal vents spew out molten rocks from still deeper in the Earth. Cooled by the ocean water and by the subterranean mountains they create ("seamounts"), these rocks are inhabited by a variety of animals, including worms

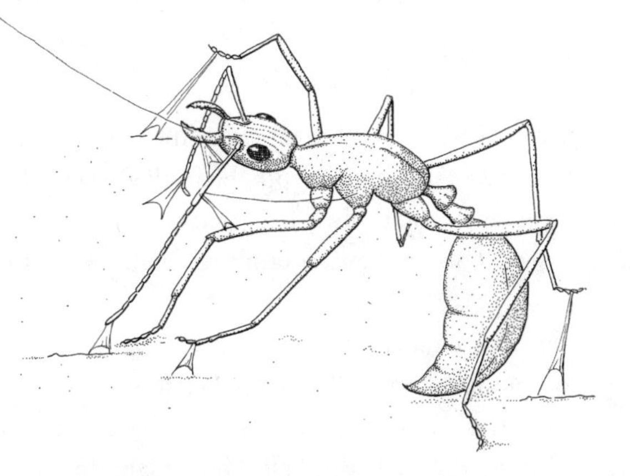

Fig. 29 A carnivorous velvet worm immobilizes an ant with instant glue. The velvet worm also has a solvent that prevents it from gluing itself.

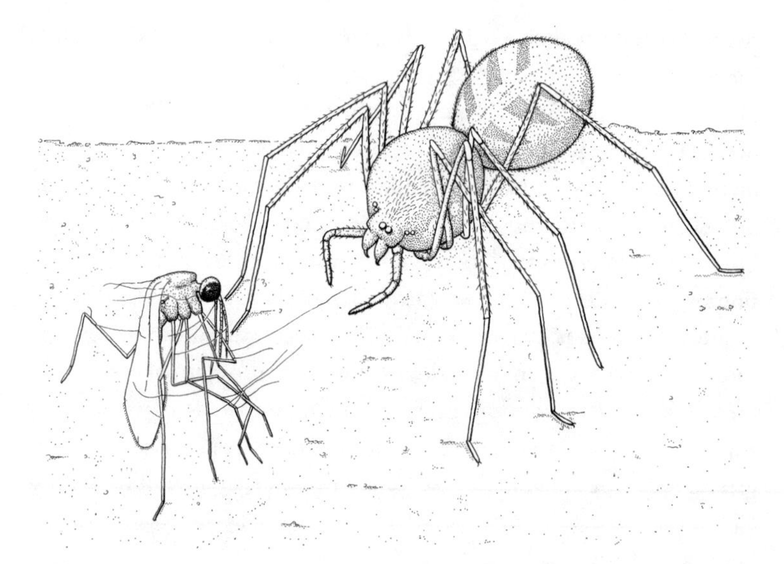

Fig. 30 The spitting spider, a fraction of a second after it entangled a mosquito with strands of instant adhesive. The capture was too fast to see even under a microscope: one moment the mosquito was about to become airborne; the next, it was doomed.

that appear to attach the tubes in which they live to the rock. What kind of glue is effective at that depth, beneath the weight of 3 kilometers of corrosive seawater, where oxygen levels are often negligible and where geysers of hot, toxic gases erupt constantly from the cauldrons below?

While we are on the subject of surviving in extreme environments, we should take a longer look at organisms that live happily in very cold or very hot locations, because they have come up with some very interesting solutions. But first, how tough can an organism be? Around the world, in soil and mosses, live common, minute organisms called water bears or, more scientifically, tardigrades (Figure 31). Each is less than half a millimeter long, and under adverse conditions they shrivel and enter a state of suspended animation. In this state they can survive boiling water, temperatures far below freezing, being placed in a vacuum, exposure to x-rays, and pressures equivalent to an ocean depth of 10,000 meters. One of the factors that allows them to do this is the ability to eliminate a high proportion of the water in their bodies without inflicting permanent damage. If we can learn exactly how they do so, we may be able to apply the knowledge to the preservation and transportation of food and other perishable materials.

In the deep-sea, hydrothermal vents mentioned a bit ago, a species of tube worm that lives there has been carefully studied—not for its glue, but for its resistance to high temperatures. Although the seawater is extremely cold at that depth, close to the vents the worms grow vigorously and often experience temperatures above 80 degrees Centigrade. How they do it is unknown but there appears to be some kind of mechanism for circulating cold water through the tube. No such luxury is possible for the bacteria inhabiting the tube or, for that matter, the waters of hot springs in places like Yellowstone National Park in the United States or Waimangu Thermal Valley in New Zealand. These tiny organisms are single celled and have evolved chemical coping mechanisms that are being intensively studied.

Anyone who has watched an egg in a frying pan will have seen the translucent protein surrounding the yolk change and become egg "white" as it heats up. This kind of change is the fate of most proteins when they are heated. They are said to become "denatured" and can no longer carry out their biological functions. When the protein is an enzyme that performs a vital function in the body, this can be a problem if the environment is too hot. However, the bacteria that inhabit hot springs have solved this problem and possess enzymes that function best at temperatures that would completely denature most others. These thermostable enzymes have many possible practical applications for the simple reason that increasing the heat in a process often makes it faster and easier. For example, dirt and stains on clothes often

Fig. 31 A group of water bears emerging from a long winter, having been protected from the cold by "antifreeze" in their bodies.

dissolve more readily in a hot wash than a cold one, and heat-stable enzymes that attack the more stubborn stains in the hot cycle of washing machines are in great demand. Another emerging application will make the manufacture of paper not only more efficient, but environmentally more friendly. Thermostable enzymes will replace the harsh chemicals used to break down wood prior to making pulp. Heat-tolerant enzymes from bacteria can be used in the process, which must be carried out in hot vats. Other enzymes are useful in the manufacture of beverages and textiles and are set for a brilliant future in various biotechnologies. Where all this will end is unknown, as discoveries are still being made. Recently a new bacterium was discovered that survives happily at 115 degrees Centigrade, or 239 degrees Fahrenheit!

At the opposite end of the temperature scale are the animals and plants that live in very cold places, on mountaintops and in polar regions. It has been known for many years that they contain familiar antifreeze chemicals such as glycerol. However, many animals have become sophisticated chemical engineers and evolved special proteins that prevent one of the most dangerous aspects of extreme cold, namely, the formation of ice inside the individual cells of the body. Such proteins have the ability to stick to ice crystals and limit their growth as the surroundings cool. They have been found in organisms ranging from plants such as the humble carrot to mussels, the eggs of insects and mites, snails, hornets, and frogs—all of which have to survive harsh winters. Still other chemicals known as cryoprotectants help organisms avoid or tolerate freezing, and these have been found in flour beetles and Antarctic fish. Ice algae actually live in sea ice, tolerating temperatures as low as −6 degrees Centigrade. While we are still learning about exactly how they work, antifreeze chemicals, some derived from fish species that live in polar regions and others from those amazing water bears, are making it far easier to store human organs for transplant operations and embryos of domesticated animals for agricultural research.

This chapter would not be complete without mentioning two further intriguing possibilities. First, the guts of many insects contain bacteria that appear to perform the same function as antifreeze proteins. Scientists have targeted certain insect pests that are particularly difficult to control, partly because they are able to survive cold winters in soils where the next season's crops will be sown. Will it be possible to infect these insects with *non*antifreeze bacteria capable of taking over the original gut populations, thereby opening the way to kill the pests with cold? The second possibility returns us to the washtub and those enzymes that remain active even in hot water. At the opposite end of the temperature scale, fat-dissolving enzymes have been discovered that operate efficiently in very cold water. Such enzymes might also make laundry easier—and cheaper. Where have they been found? On the corpses of whales in frigid ocean depths. The research continues.

A short while ago we suggested that the glues made by mollusks, tube worms, and their relatives are of commercial interest because they are so effective. In a strange twist of fate, we now want to talk about these animals from the point of view of finding ways to unglue them! The reason is straightforward: these animals are very welcome when living in their natural habitats, but when they attach themselves to docks, piers, wharves, oil rigs, yachts, cruise ships, oil tankers, and aircraft carriers, they become real pests. (We are not going to get into the argument about who was there first. Of course the animals were, but the fact remains that in some locations they are a real nuisance.) For example, a ship carrying a load of marine animals stuck to its hull is much slower and costs much more in fuel and time than a ship with a clean, slick hull. The animals and plants that stick to ships and piers and other human undersea structures are known as fouling organisms. The search is on for antifouling chemicals.

In fact, such man-made chemicals exist, but they have run into serious trouble because they contain copper or tin and kill all kinds of wildlife besides the fouling organisms. Fortunately this is no longer acceptable, and environmentally friendly chemicals are being sought. Enlightened chemists have been looking beyond the chemical store in the laboratory to the chemical store in the sea. In particular, they have put on their diving suits to search out living species that appear to avoid being encrusted with fouling organisms. They have found a lot of these species, mostly in plants and animals that cannot escape because they are themselves fixed to the spot and unable to move away. These organisms include corals, sea grasses, sponges, seaweeds, and curious plantlike bryozoans (moss animals), all of which have yielded promising antifouling chemicals. Even the bacteria are in on the act: a species that inhabits the skin of sea squirts produces strong repellents. Of course, this is a highly incestuous battle; each species must attach somewhere, and that "somewhere" is often another sedentary species! Although barnacles resist settling by other larvae, they may be quite happy attempting to settle on a robust seaweed that is producing chemicals to repel settlers!

A variety of antifouling chemicals derived from living organisms are nearing production. However, the trick is to find a chemical that is strong enough to do the job but not so strong that it kills nontarget species and wildlife. It must be capable of being manufactured in large quantities and therefore not too expensive, and it must be able to be incorporated into a long-lasting paint. All these are difficult problems, but the possibilities are enormous. The available choices include about five thousand species of bryozoans and many more thousands of corals, sponges, and seaweeds.

In medical research it is often desirable to treat a particular group of cells and then see what happens. To do this, scientists need some sort of signal that the particular change they are hoping for has occurred. The use of color is ideal, whether it can be seen with the naked eye or has to be activated with an instrument such as a laser. The question has therefore arisen, where have useful colors evolved? The answers are both varied and dramatic. Among the most promising is the luminescent chemical coelenterazine from a species of jellyfish. Both an inspiration and a blueprint, this chemical has many of the desired properties of a signal but is difficult to extract from the host. However, natural-product chemists have analyzed it and now know how to synthesize it in quantities sufficient and affordable for research purposes. Further, they are using the blueprint of the jellyfish molecule to make other color signals that may generate extraordinary new opportunities. For example, crop plants may be engineered to incorporate a molecule that changes color if the crop is short of water or under attack from viruses. Other useful color molecules, or pigments, come from fireflies, cyanobacteria, and one of the googly-eyed, toothy fish from pitch-black ocean depths where luminescent colors are the only visual means of detecting mates or prey.

Researchers look forward to tracking several events at once through isolation of the genes responsible for the dramatic colors of the coral reef. The genes that produce the color proteins are inserted next to the gene under examination; when that gene is active, it switches on the color. In addition to green, red, yellow, and blue are now available. Why these colors evolved in corals is anyone's guess, but some researchers suspect that they may increase the coral's resistance to the damaging effects of the sun's ultraviolet light. In other words, they act as sunblocks.

Natural chemicals not only constitute one of the most fascinating topics in the sciences, they have a proven record of success in generating commercially useful products. The examples we have discussed appear to be just the beginning. Hundreds of thousands of species have already solved their problems of repelling enemies, gluing themselves to safe havens, preventing others from gluing themselves to sites already occupied, surviving in extreme cold, or surviving in extreme heat. The examples together include plant, animal, and microbial species—many known to science for centuries, others only recently discovered. Various industries are already deeply involved in exploring this chemical cornucopia, and a great many more products await discovery.

ELEVEN
Blueprints and Inspiration from the Wild

The first two paragraphs of this chapter belong to two scientists, writing together with their teams in two of the top scientific journals in the world. The first is B. L. Smith, in *Nature*: "Natural materials are renowned for their strength and toughness. Spider dragline silk has a breakage energy per unit weight two orders of magnitude greater than high tensile steel, and is representative of many other strong natural fibres. The abalone shell, a composite of calcium carbonate plates sandwiched between organic material, is 3000 times more fracture resistant than a single crystal of the pure material."

The second is A. H. Heuer, who wrote in *Science*: "By adapting biological principles, materials scientists are attempting to produce novel materials. To date, neither the elegance of the biomineral assembly mechanisms nor the intricate composite microarchitectures have been duplicated by nonbiological processing. However, substantial progress has been made in the understanding of how biomineralization occurs, and the first steps are now being taken to exploit the basic principles involved." This passage was written in 1992, and the field has never looked back.

Over hundreds of millions of years the Earth has gradually become home to millions of species of animals, plants, and microbes. During this time they have encountered and solved many of the same problems humankind faces today, including finding or constructing a place to live, harvesting food, and fending off disease. The big difference between humans and most other organisms on Earth is that while they have been around for a very long time, we are relatively new to the planet and are still struggling to solve these problems. In this chapter we continue the theme that we may learn how to solve our problems by studying how other species solve theirs. In the process, we look at naturally evolved solutions that point the way to new industries or to dramatic advances in established ones.

Biomaterials

Looking for useful materials in nature is one of the oldest of human occupations. Our ancestors once bound bones or sharp stones to sticks with vines or animal sinews to make weapons for hunting. Even now, anyone who has been beachcombing, made a campfire, or rummaged through a garage sale, understands the pleasure in finding useful natural materials such as coral, tinder, cane, or mother of pearl. Moreover, our long history of interest in and dependence on natural materials has flowered into many industrial products and enterprises based on the simple idea that a material needed by humans may well have an analogue in nature.

Mollusks that live in the sea make amazing and beautiful shells. To do so, they extract minerals (mostly calcite or aragonite) from the water that surrounds them and construct layers of crystals embedded in proteins and other organic compounds—somewhat like bricks and mortar. This material is lightweight, yet tough enough to withstand the immense pressure of waves or depth and the attacks of predators. At the same time, it is sufficiently flexible and dynamic to allow the animal inside to grow and to move. The structures perfected by these animals have been brought into laboratories around the world and examined in great detail. For scientists and engineers are searching for materials with the peculiar combination of strength, hardness, and flexibility that has evolved in the shells of many mollusk species. Consider the apparently contradictory demands of a material such as flexible concrete, which has been modeled on the structure of the mollusk shell. The layered structure of the few species that have thus far been analyzed is also inspiring new ceramics, some designed for the manufacture of turbine blades in jet engines that have to withstand extreme levels of stress, heat, and corrosion.

The processes whereby mollusks build shell, mother of pearl, and similar materials are collectively known as biomineralization. There are many ways of breaking down these processes into segments small enough to study and understand. The molluskan radula has been particularly inspiring (Figure 32). This organ is best described as a tongue, but it is usually lined with extremely hard teeth. It is adapted for many different ways of obtaining food, including the type of herbivory that involves scraping algae off rocks (and the sides of aquariums) and a brand of carnivory in which predatory mollusks drill holes through the shells of their prey, which are also mollusks. The radula is a continuously growing organ, as its tip wears away with use. The net effect is a conveyor belt with the young, soft, unmineralized teeth at one end and the older, harder, biomineralized representatives at the other. Analysis of these teeth from their first formation through every stage to maturity has enabled researchers to learn much about biomineralization and how it may be engineered for a variety of purposes. The materials themselves have yielded new insights into the manufacture of industrial ceramics, and the radula has become a model for developments in two somewhat equivalent human activities, dredging and mineral processing!

Although this type of research is widely known as biomimicry, the word applies equally to most of the other kinds of research and development described in these chapters. The common thread is the question that pops up repeatedly: Where might we expect the product, in this case the material, to have evolved? Well over one hundred thousand species of mollusks have been described. Most have a

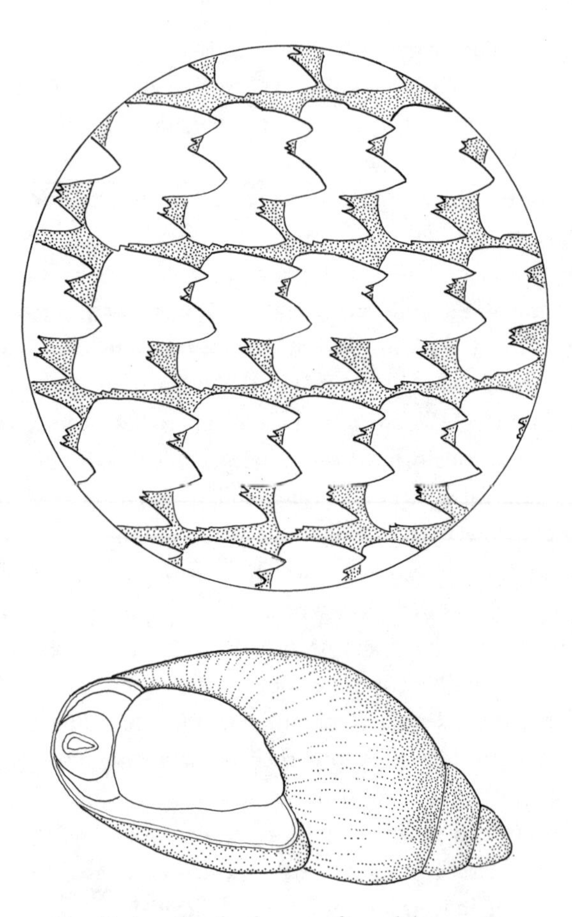

Fig. 32 A snail and a close-up of its radula, a rasping tongue made of materials interesting to engineers.

shell, either outside the body (snails, clams, oysters) or inside (cuttlefish, squid). Biologists reckon that there are many more species to be found. The shells are adaptations to life in many different places, including hot deserts, rain forests (many species live exclusively either on the floor or in the canopy), freshwater, brackish water, in the sea where waves batter the rocks, or at great depths often deep in mud. Last but not least are the mollusk species with highly specialized shells capable of boring into timber or rock. The industries involved with the development and manufacture of mollusk-inspired materials are making major advances. Mollusk diversity is a vast resource base for materials scientists and engineers.

In addition, a variety of other natural materials are under scrutiny for their potential. Chitin, the material of the exoskeletons (the hard external parts) of a wide variety of animals such as insects and crabs, is being considered for surgical sutures and dressings. Not only is it strong and protective, but the body naturally absorbs it, eliminating the difficult and painful steps of removal from wounds. Hedgehog spines, which contain soft, honeycomb-like materials that enable them to bend without buckling, have inspired lightweight wheels in which the tires are replaced with dense arrays of spines that make magnificent shock absorbers. Plant materials offer many opportunities for new materials or new uses for old ones. The anatomy of wood with its long, tough, overlapping cells is the model for new, much lighter construction materials. The ancient crop known as kenaf, a relative of okra and hollyhock, is now the basic resource for a company using it to manufacture interior panels for car doors, seat backs, and interior trim—first steps toward a truly recyclable automobile.

Silk

Silk is such a common and well-loved material that it fails to strike most people that we are totally dependent upon insects, spiders, and their immediate relatives for its production. To be sure, man-made imitations abound; but if the real thing is needed, we turn to the natural-silk factories, the best-known of which are silk moths. Most production in the past has been for the manufacture of clothing. However, silk has many properties that are highly sought after. Laboratories the world over are studying this remarkable material for application in a wide variety of industries.

The garden spider that spins the cartwheel type of web does so because she has a particular problem to solve. (It is always a female; the males merely lurk nearby waiting for a chance to mate.) Her trap, the equivalent of a fishing net, hangs in air rather than in water. So without the buoyancy provided by water it must be light in weight. Yet it cannot be weak, because the spider's prey is often a large insect such as a beetle with a very tough exterior, flying at high speed. There is another problem: the spider's mouth is so tiny that its meals must be fluid. To convert solid prey into a liquid meal, the spider's jaws create a small hole in the freshly killed prey through which digestive enzymes are injected. The carcass becomes an external stomach, its contents a mixture of enzymes and semidigested organs and muscles. When the contents have been fully broken down into nutritious fluids, the spider sucks them back into its own body. As a result, the silk must be able to absorb large amounts of energy so that the prey does not smash into pieces when it hits the web. Otherwise, the spider would not be able to set up its external digestive system.

Now we see the spider's special problem. It must create a net made from a fiber that is light in weight but very strong, and elastic enough to absorb all the energy of a high-speed impact (Figure 33). Such a fiber is a dream come true for some industries. It turns out that spider silk is as strong as steel, but at the same time extremely elastic and of course ultralightweight. Before it snaps, a thread of spider silk absorbs roughly a hundred times more energy than a steel filament of the same size. Even though such properties have attracted much attention, they can be matched or exceeded by synthetic fibers—fused silica, graphite, beryllium oxide, and the material marketed as Kevlar. Thus, the issue is not simply that spider silk is an astonishing material, but how exactly it may be a wild solution.

Suggestions for possible uses of man-made versions of spider silk

Fig. 33 The net-casting spider makes several different kinds of silk, including the trap held by its four front legs. This silk has to expand and contract in a split second as the spider attempts to snare a passing moth.

include superior bullet-proof vests or body armor, light but very strong safety nets to catch the top guns that overshoot the flight decks on aircraft carriers, even cables of the suspension bridges of the future. Other suggestions recognized that silk's additional properties of being waterproof and nonallergenic, when added to its strength and elasticity, may be useful in medical technology for artificial tendons, burn dressings, and strong, ultrathin sutures. Whether for these goals or for new materials for fishing nets or holding six-packs together, the commercial development of spider silk is forging ahead and attracting serious investment.

In one of the major advances, the genes responsible for production of the silk proteins have been identified, isolated, and transferred to bacteria. In turn, the bacteria have been grown in large quantities and have produced silk proteins in bulk. The troublesome problem is that bacteria cannot spin. The technology that has evolved in the spider's spinnerets, the technology that transforms the proteins from a soluble form inside the body into long, ultrathin, waterproof threads outside the body, has yet to be efficiently mimicked in the factory. In one approach the silk proteins are extruded through tiny tubes into solutions where threads form as the proteins crystallize. Another company has inserted the silk-protein genes into goat udders, to be harvested with the milk. Several laboratories are taking spider spinnerets apart, molecule by molecule, in the belief that it is from the original factory that the wild solution will come. Scientists are enthusiastic about these developments, not only because new silk-like fibers will result, but also because the gentle anatomy and biochemistry of insect organs point the way to less polluting chemical technologies.

All of this exciting research has focused on just one or two kinds of spiders, but thousands of species lurk in every corner of the world and are an immense and largely unexplored resource base for the fiber industry. Different spiders make different kinds of silk that vary in elasticity, strength, and degree of stickiness (see Figure 34). A huge variety of silk-producing technologies is reflected in the fascinating array of webs. But spiders are not the only silk producers. In addition, there are moths and butterflies, pseudoscorpions, mayflies, bugs, lacewings, and many kinds of beetles. On warm summer evenings "dance flies" swoop to and fro above ponds and streams. Examination of these shimmering swarms reveals courting males, offering silk-wrapped

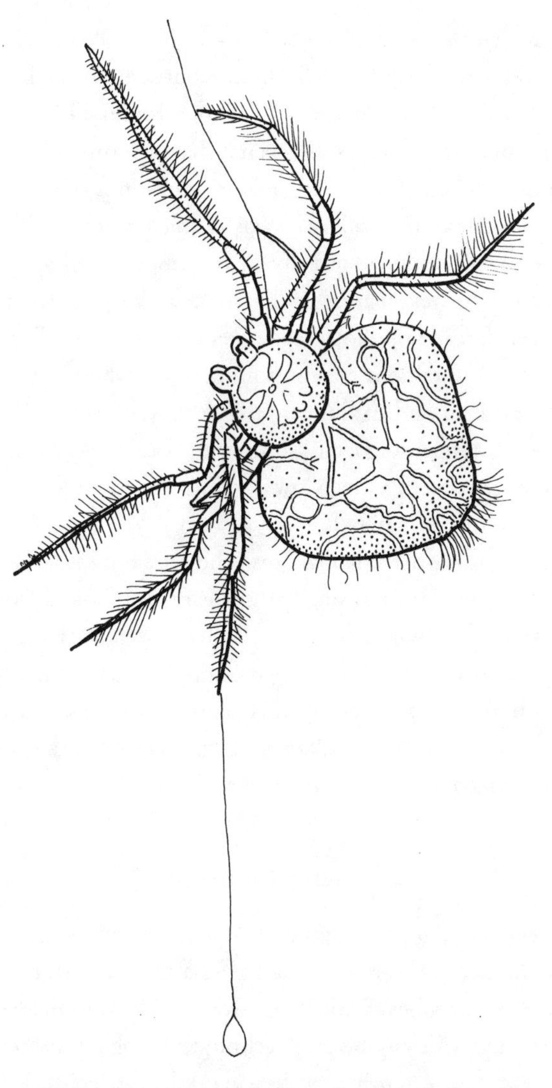

Fig. 34 The bolas spider spins a short length of special silk with a drop of glue at the tip. When the spider detects the vibration of a flying insect, the trap line is whirled beneath its body.

gifts to the females. The silk emerges from the swollen forelegs of the males. Most often the gifts are fresh prey, but sometimes the silk conceals a worthless, inedible scrap. To emphasize how little is known, this year a silk producer was discovered that has stunned the experts. (We should explain that the larvae of various kinds of ants spin silk, and the construction of silk nests has enabled many of them to become highly successful, especially in the tropics. Perhaps the most spectacular is the weaver ant, which holds a larva in its jaws and uses it rather like a darning needle to sew leaves together.) The discovery occurred in a mountainous area of South Africa, where entomologists for the first time found an adult ant that makes silk. A totally different species, it has special spigots just below its mouth and special brushes on its front legs that together allow it to line and repair its nest tunnels with the waterproof material.

Then there are the Tanaidaceans. These are not the latest enemies of Starship Enterprise, but small crustaceans that look like miniature lobsters and spin silk in seawater at enormous depths (Figure 35). Finally, other insects such as sawflies produce a variety of fibers similar to silk with interesting physical and chemical properties. Overall, the variety of silk-producing animals and silk-manufacturing technologies that awaits our curiosity is immense.

Bacterial Industries

In the previous section we mentioned the transfer of the silk gene into bacteria and the cultivation of the bacteria in huge vats for the production of silk proteins. While there were problems in that particular case, current genetic engineering techniques recognize bacteria as superb protein-manufacturing machines and they are being deployed in a host of different ways. Proteins are a major component of the human body and are therefore of great importance in medicine. Research is also focusing on artificial proteins that might incorporate in

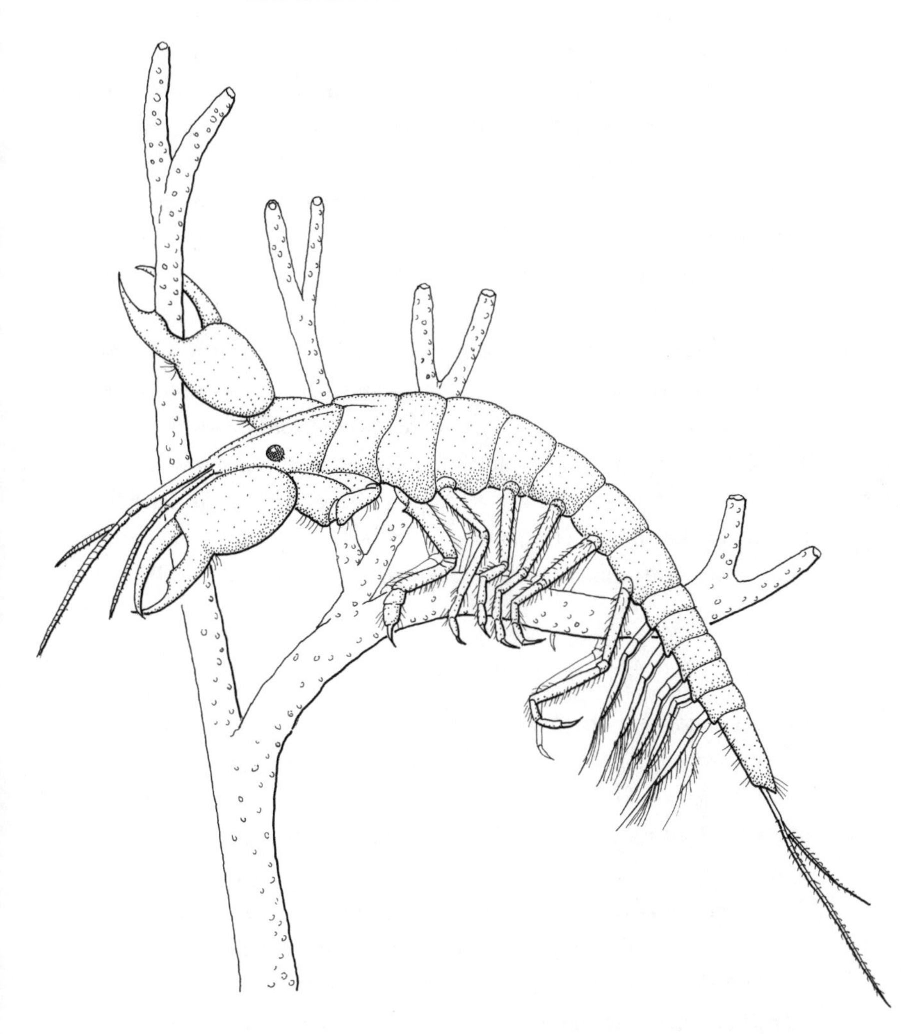

Fig. 35 Tanaidaceans are small, shrimplike crustaceans that occur throughout the oceans of the world. Some live in deep trenches and spin silk to make their tunnels. What kind of silk has evolved to withstand such rugged conditions?

a single molecule the advantages of both natural and laboratory-made proteins. For example, certain natural proteins bind cells together, and these are critical in the healing of wounds, especially when finding enough tissue depends on the application of cells grown in tissue cultures. Artificial proteins can be grown in bacteria to combine the superior adhesive properties of natural proteins with the superior stability of artificial ones.

Medicine is not the only field to take advantage of the bacterial factory; the microbes are also being used for the manufacture of biodegradable plastics. While biotechnology and genetic engineering are not the focus of this book (they require a whole volume unto themselves), we do need to point out that the engines of these growing industries—bacteria—are only a fraction of a millimeter in length. Indeed, to place them in the context of our book, they answer the question, where have sophisticated, precision protein factories evolved?

Robotics

The humanoid robots constructed by mad scientists that stalked innocent townsfolk in black-and-white horror movies would have been very inefficient. For starters, while walking on two legs facilitates many kinds of movements (ask any athlete), much of the processing power in the robot's on-board computer would be taken up in keeping the contraption from falling over. Lately, serious robot engineers have searched animal diversity for models that inspire design primarily for stability, but also for simplicity and maneuverability. To this end, many have turned to the arthropods, investigating various insects and their relatives, including the arachnids, crustacea, and the centipedes and millipedes. The word "arthropod" means jointed legs, and these animals have been perfecting their legs and the locomotion generated thereby for hundreds of millions of years.

Insects with their six legs are very stable. Movement generally involves keeping the front and back legs of one side touching the ground at the same time as the middle leg of the other side. Six legs are also more stable than two or four when the animal, or robot, is carrying a large object. Insects in particular have inspired the design of various kinds of six-legged robots required to forage by themselves in dangerous situations (searching for terrorist bombs, exploring damaged nuclear power plants and chemical dumps, or reporting on earthquakes and volcanic eruptions). Robots involved in space exploration may have similar six-legged designs. Understanding the ways insects move is not easy and requires teams of entomologists, engineers, and mathematicians to break each movement into its components and then recreate them with electronic hardware and computer software. Even then, research into insect movement suggests that robotic engineers will find it difficult to mimic the speed and agility of the animals themselves. Cockroaches, for example, can scoot one meter in a second, in the dark, while their finely tuned sensors and muscle systems allow them to make up to twenty-five twists and turns in that brief time. Nevertheless, engineers are building six-legged robots with controls inspired by the insect nervous system and by the complex sensors in insect appendages that control posture and inform the central processing unit of their position.

Insects are not the only animals providing inspiration for new robotic designs. Crabs are being mimicked to produce robots with long strides made possible by their relatively simple legs and by the fact that they walk sideways. Millipedes with many pairs of legs arranged along their tubelike bodies are being investigated for robots to carry heavy weights in cramped conditions, where much twisting and turning is required. Worm locomotion may provide the key to constructing robots that can explore the corners of our digestive system inaccessible to conventional endoscopes; and by understanding the way snakes move, robots may be constructed to handle terrain inaccessible to wheels or legs. Fish, especially fast swimmers such as tuna, are the models for underwater robots that will map the ocean floor, detect the extent of oil spills, and measure oceanic temperature changes. The fish shape, which provides efficient movement and speed, is of special interest to these engineers so that the robot can be designed to remain active for long periods without refueling.

Finally, research into the amazing flight of flies, moths, and dragonflies is being pursued. These insects are well known for their speed and agility in the air, using a form of aerodynamic movement that until recently has defied analysis. By placing the animals in wind tunnels through which nontoxic smoke flows and photographing their flight in slow motion, researchers have been able to see not only how the wings move and bend, but what they do to the air to create lift and maneuverability. One current area of research is the building of tiny aircraft capable of rapid, weaving flight, with on-board cameras for reconnaissance in tight, enclosed situations. These craft will be based on the mechanics of insect flight and may also incorporate analogues of the nervous and visual systems evolved in flies to cope with their rapid, twisting flight patterns.

Two engineers involved with robot fish and moth flight, Michael and George Triantafyllou, wrote a few sentences that capture the theme of this chapter: "The more sophisticated our robotic tuna becomes, the more admiration we have for its flesh and blood model. Aware we will never match the perfection of design of the living creature, we strive instead to uncover natural, useful mechanisms optimized by millions of years of evolution." And Isao Shimoyama, speaking of wild moths, says: "Insects are ideal models. Their flight mechanisms have already been tried and tested through millions of years of natural selection." We couldn't have said it better ourselves.

Fire and Smoke Detectors

Where would you expect to find a naturally evolved smoke detector? Although an answer seems improbable, research has provided one. Once again, the database is natural history, the storehouse is biological diversity, and the skills required are those of the evolutionary biologist, the ecologist, and the natural-products chemist.

We may begin by asking which animals are particularly susceptible to fire? The answer is that all animals are in grave danger when exposed to this ancient enemy, but perhaps those that are small or confined—as in a web or cocoon—or have limited mobility are especially vulnerable. However, this description covers such a wide variety of animals that we are not helped very much to focus on an answer. In the 1994 fire that consumed more than 90 percent of the oldest national park in the world, Royal National Park just south of Sydney, some survivors were able to hold out on isolated rocky outcrops. There was no suggestion that they possessed any kind of fire detection system; they were just lucky. Because all animals are vulnerable to fire and because survival appears to be a matter of chance, it seems that this line of argument will not take us very far.

At this point, we go back to our database of natural history and our storehouse of animals, plants, and microbes and turn the question around: Are there animals that have fire-detection systems because they actually *need* fire? While this possibility appears highly unlikely, one kind of natural historian, the beetle entomologist, could reply: "Of course there are animals that seek fire; some buprestid beetles lay their eggs only in trees recently killed by it. They must be able to find these trees, so they may have evolved smoke detectors." This sort of knowledge is both valuable and scarce. Few people know what buprestid beetles are. In fact, their common name is jewel beetles because of their glossy, iridescent colors. To entomologists some jewel beetles are also known as fire beetles, because they fly toward forest fires and, once the trees have cooled sufficiently, lay their eggs in the cooling but dead wood—the food preferred by the larvae. Why do they prefer freshly killed wood? We know that wood is nutritious—hundreds of species of termites and other wood-boring insects demonstrate that. So maybe the reason for colonizing a tree that has just been effectively heat sterilized is that the fire beetle larvae have no competition from other species. It is also very likely that any chemical defenses in the tree have been neutralized by the high temperatures of the fire.

The beetles are equipped with two kinds of extremely sensitive sensors (Figure 36). The thorax bears a pair of organs, each of which is located at the base of a middle leg, which is held aloft when in flight as though to expose it to the air. These sensors are extremely sensitive to infrared and enable the beetles to detect fires over large distances. Anecdotes suggest that they can find a fire as much as 50 kilometers away. This amazing ability is augmented by sensors on the antennae that detect smoke. Moreover, there is experimental evidence that the beetles can tell the difference between the smoke of their preferred trees (pines) and all the others. This exciting research continues, as it is unclear exactly how the beetles integrate the two sensors. Possibly they use their infrared detectors to home in on a fire from a distance, then, when close up, sample the smoke to determine if pine trees are involved. Modern infrared and smoke detectors are also extremely sensitive, so commercial interest in the beetles may focus on the relative costs of conventional technology versus beetle technology. If technologies derived from the mechanisms discovered in the beetles turned out to be cheaper, fire detection could be more affordable and hence become more widespread. Think of the advantages of widespread, cheap fire detection in situations ranging from production forests to apartment blocks to warehouses!

Fig. 36 These fire beetles have extremely sensitive infrared detectors where the middle pair of legs emerges from the body, and smoke detectors in their antennae. Because they fly with these legs elevated so that the detectors are free from obstructions, they can detect fires many kilometers away.

Air-Conditioning

Termite mounds are among the most common sights in tropical and subtropical grasslands. Nature documentaries often show them standing like so many gravestones on the brown plains, their dusty towers shimmering in the sun. How can their inhabitants survive the relentless baking heat? The answer is that the mounds are air-conditioned.

Although the outside air is often extremely hot, the inside temperature is constant and much cooler.

The idea behind termite air-conditioning is simple. The mound itself is a tower, full of tunnels that are, of course, full of air. As the sun raises the temperature of the mound during the day, the air within warms up and rises, eventually escaping to the outside through tiny pores in the mound walls, which are made of compacted soil. Any breezes around the mound help this flow of warm air away from the colony. To complete the circulation, the colony requires that cool air flow from deep underground, where the heat of the sun does not reach. Many termite mounds do indeed have underground tunnels and chambers, including basement caverns containing cool air from the lower layers of soil. The bulk of the colony inhabits the lower layers of the mound, but the activities of millions of termites continuously warm it (imagine how much heat a million people would generate in an enclosed space). The air rises into the tower above, eventually escaping through its porous walls. Termite air-conditioning thus uses soil-cooled air to keep the insects comfortable and, as they warm the air, disposes of it up a porous chimney. Their air-conditioning therefore takes advantage of a constant and well-known law of nature: warm air rises. Not only is this solution ingenious, it is free.

At this point let us pause for a moment to ask about the costs of air-conditioning for the termites. The manufacture of the mound and tower is undoubtedly the principal cost. The excavation of the tunnels and chambers below ground and the construction of the towers above ground are achieved grain by grain, each particle of soil or sand carried in the tiny jaws of millions of termite workers. In those species that build large towers, some as high as two cars piled on top of each other, the process may take years. Some mounds may last for centuries. The cost of mound building in terms of the initial investment of energy is therefore large. On the other hand, the costs of maintaining the system are minimal; only minor equivalents of electricity bills come to the termite colonies; it is the natural movement of air that keeps the colony cool, not the products of large, expensive utility plants.

It may seem unlikely that termite mounds could act as blueprints for architects and construction engineers, but that is exactly what they have done. With the costs of electricity rising in terms of both dollars and climate-altering pollutants, these professionals have turned to the humble insects for inspiration and have designed and constructed public buildings using the principles of termite air-conditioning. Cooling breezes are distributed through complex ducting that leads to glass or brick towers, which may differ architecturally but are basically the equivalent of termite chimneys. By the time the air reaches the towers, it has absorbed the unwanted heat from the offices, laboratories, or theaters within the building and thus rises and escapes to the outside, just as it does from a termite mound.

Energy

This final section, our colleagues insist, is the most speculative. Their insistence is based on the natural reluctance of scientists to look too far ahead because of their intimate knowledge of the pitfalls and false starts of scientific research. Yet several groups of researchers around the world have gone ahead and asked the deceptively simple question, Can plants show us a better way of harnessing the energy of the sun? Or, to put it another way, is it possible to make an artificial leaf? We should emphasize right away that we are not advocating the replacement of real leaves with artificial ones! On the contrary, we support the urgent conservation goals of restoring the leaf-laden forests, grasslands, mangroves, and sea-grass beds that have been damaged all over the world. Rather, the idea of an artificial leaf is aimed at reducing reliance on fossil fuels by generating energy more efficiently and in ways that reduce harm to the environment, especially the atmosphere. Fuels such as oil, coal, and gas are derived from ancient plants whose leaves trapped sunlight millions of years ago. They are in this sense fossil sunlight, but their combustion generates widespread pollution. So we turn to a third way of posing our question: Can we trap sunlight today in a way that seriously reduces the pollution released while we generate energy?

To approach the question from the perspective of this book, we need to look for a naturally evolved mechanism that enables an organism to absorb and utilize solar energy. This mechanism is photosynthesis, which occurs in a wide variety of organisms. Photosynthesis means "assembling with light" and it basically involves the conversion of carbon dioxide and water into organic compounds required for growth, using the energy in sunlight. Light energy drives the process and is converted into chemical energy by highly specialized molecules called pigments.

The process probably first evolved in the phototrophic bacteria that include the green or purple sulfur bacteria, which often turn mud those colors. Many people first encounter sulfur bacteria when walking on mudflats, where the smell of rotten eggs taints the air. Unlikely as it may seem, these bacteria actually require hydrogen sulfide, the gas that causes this smell, as a fuel for photosynthesis. The bacteria are abundant and include specialists for many habitats (such as those smelly mudflats) that are unappetizing to humans. The number of species involved is likely to be huge, but right now no one knows how many there are.

Other forms of photosynthesis occur in the cyanobacteria, the algae, and the flowering plants, all of which harvest solar energy for the manufacture of food. As in the phototrophic bacteria, carbohydrates are assembled from carbon dioxide—but for these organisms the raw ingredients are water and inorganic salts. Again the source of energy is sunlight, but the pigment is one of the familiar green-colored chlorophylls. The cyanobacteria have been around for a staggering 3 billion years and can still be common today where conditions meet their requirements. The algae appeared later, diversifying into hundreds of thousands of red, green, and brown species. New species of algae are being found all the time, and algologists are certain that many more await discovery in all parts of the world. The flowering plants appeared about 130 million years ago, and there are perhaps 350,000 species, although the total number that inhabit Earth remains unknown. Over these immense periods of time, a variety of photosynthetic structures and pigments have evolved.

Scientists have been studying photosynthesis in detail for many years. The truth is that while we know a lot about it, the goal of creating artificial photosynthesis with anything like the flair of plants still seems a long way off. The chemical interactions that take place within every photosynthesizing cell not only depend on a highly organized sequence of events, but are mediated by the subtle structure of tiny organs known as chloroplasts, which consist of many layers of delicate membranes. The chemicals and organs have evolved over millions of years and are impressive for their efficiency and elegance. Engineers need to step back and decide whether they are going to use the chloroplast as the inspiration for an artificial mimic, or as the blueprint to reconstruct what nature has made. Both approaches have been adopted. Among the mimic teams, chemists are having some success experimenting with layers of various metals, including rhodium, iridium, and platinum. Meanwhile, the blueprint teams are studying the various pigments and subcellular structures that abound in nature and have made progress with simple cell-like devices called liposomes, which resemble the membranes needed for critical photosynthetic chemical reactions. Everyone agrees that the blueprints are available in nature, but how to emulate them is one of the most complicated tasks facing modern plant biology and chemistry.

TWELVE

Wild Medicine

Medical knowledge, accumulated over thousands of years, today is exploding as a result of the massive sums pumped into medical research. Nevertheless, many human ailments and diseases persist and new medical problems, many generated by modern life styles, add to the demand for cures and treatments. In Chapter 9 we saw how the exploration of biological diversity may lead to the discovery of novel antibiotics. In this chapter we will go much further, into a variety of areas of medicine, and ask if the solutions we are seeking have already evolved in the wild.

Rain Forests, Coral Reefs, and Beyond

The species that inhabit rain forests and coral reefs are well known as reservoirs of medicinally useful chemicals. The quinine plant of Peru, which produces a treatment for malaria, and the rosy periwinkle from Madagascar, which has yielded potent anticancer drugs, are two examples from rain forests. The Moreton Bay chestnut from the Australian tropics (Figure 37) not only generates timber fine enough to be used for the speaker's chair in the British Parliament, but also contains an intriguing chemical, castanospermine, that has been investigated for the treatment of AIDS.

Coral reef products include anti-inflammatory, antiviral, and antitumor compounds isolated from a variety of invertebrates—especially

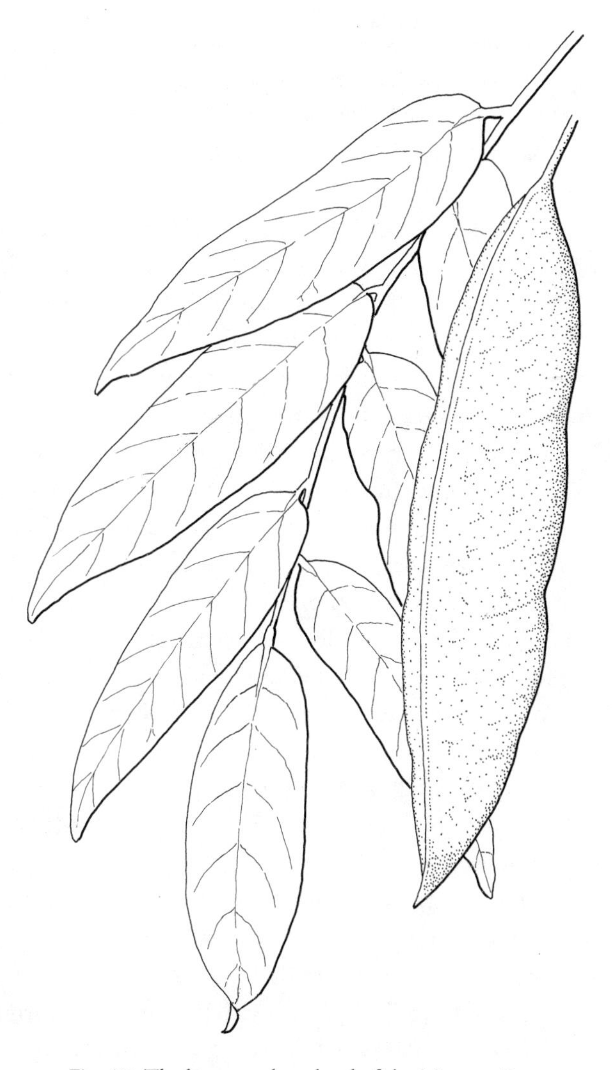

Fig. 37 The leaves and seedpod of the Moreton Bay chestnut, which contains chemicals of major importance to cancer research.

sponges, corals, and sea squirts. In the case of an anticancer compound discovered in one particular species of sea squirt, the great frustration was the infinitesimal amounts in each animal, never enough for the required testing. However, by applying a little ecological thinking, scientists may be able to solve the problem. The sea squirt has a predator, the tiger worm, which does not digest the useful chemical but stores it in its own tissues. As each worm eats a hearty diet of sea squirts, it stores a lot of chemical. So it is just possible that by farming both the sea squirt and its predator, the predator can be harvested for the drug. The goal of farming the animals using aquaculture techniques will decrease the pressure on the natural populations.

You may well ask why the tiger worm stores the chemical in the first place. Does it suffer from cancer? The most likely explanation is that the chemical has evolved as a toxin to protect the sea squirt; and the worm, like many other predators that eat poisonous prey, cleverly retains the toxins for its own defense. Corals also make chemical sunscreens that protect their tiny, jelly bodies from ultraviolet rays, which penetrate to remarkable depths under the fierce tropical sun. These are about to become commercially important as they have the added advantage of being hypoallergenic. While we tend to think of sunscreens in terms of rubbing them on our skin when we are exposed to the sun, their major uses are industrial, as for the protection of plastics and paints from weathering.

Estimates of the value of drugs from rain forests and coral reefs vary greatly. Some are based chiefly on sales (wholesale or retail), while others attempt to include the social aspects such as the value of lives saved or careers restored. Despite this variation, individual estimates are truly impressive. Drugs derived from the rosy periwinkle have worldwide sales of approximately $160 million per year. At least forty-seven major drugs have come from rain-forest plant species, and it is estimated they have an accumulated net worth of $147 billion. The retail value of all plant-based pharmaceuticals in the United States during

1990 was $15.5 billion. Had this estimate included the value of all the lives saved and the pain alleviated, the estimate would have been far higher.

Local or indigenous peoples form a significant source of knowledge, and there has been enormous interest in folk medicine. However, indigenous peoples may be unaware of the vast profits that can be made from their medicinal knowledge or unenlightened about how to negotiate a share in those profits. The result is serious ethical issues about the use of this knowledge. Recognition of the differences between large multinational companies and local peoples and cultures with respect to economic, legal, and political power has led to some major improvements in the conduct of this kind of business. There has also been some international agreement as to who owns the knowledge and the species involved, and how the benefits may be shared; but much remains to be done. Perhaps the best-known agreement is between the Instituto Nacional de Biodiversidad in Costa Rica and the pharmaceutical giant Merck, which has funded bioprospecting for plant and insect species of medicinal value in the rain forests of that Central American nation. The hope is that new drugs will be found that are effective against cancer and diseases caused by bacteria, fungi, or nematodes. Other pharmaceutical companies have reached bioprospecting agreements with Native Americans and with local peoples in the state of Kerala, in southern India.

An interesting approach to rain-forest plants has been developed by botanists investigating their uses by indigenous peoples. The researchers first reach agreement with their hosts on how the knowledge gained will be used. Then, in the forest, they mark out a standard study plot, generally a hectare, and work their way slowly through it, identifying every tree and asking their hosts how each is used. Usually trees can contribute in many ways to the local population: as food, construction materials, dyes, decoration, medicines, or the source of poisons for hunting.

The results often challenge the common perception that timber is the chief use of forests. For example, a major study of tree use by several different Amazonian groups showed that the number of species varied between 70 and 119 per hectare plot, and the proportion that were useful to the local people varied between 49 percent and 79 percent. Of these, between 7 and 35 percent were used medicinally. Considering that the survey included just trees and not any other types of plants or any of the animals, the results suggest a very large reservoir of local medicinal knowledge and many medicinal resources in forest biodiversity. Another useful outcome of the same study was identification of the plant families that had the most uses. The most serviceable plant family turned out to be the palms.

In Thailand, researchers are attempting to place a dollar value on the medicinal plants used by local peoples. The value of the plants collected and used in one national park is compared to the costs the villagers would have incurred had they been forced to buy drugs from a store. These economies, together with the savings through being able to continue their normal activities, soon climb to many millions of dollars.

Local knowledge of medicinal plants has the potential to save many lives and, as a bonus, to help the local economy. An example comes from a pretty West African tree called millettia. Many of its relatives are ground up and thrown into rivers to stupefy fish so that they will be easy to catch. However, when the seeds of this particular species have been tried, they leave fish unharmed but kill water snails. This simple observation has been developed by researchers into a possible treatment for one of the world's biggest killers, schistosomiasis (or bilharzia). The disease is caused by a tiny worm that requires both the human body and the body of a freshwater snail to complete its life cycle. The snails release large numbers of worms in their feces and people become infected when they enter the water to bathe, fish, or plant crops. The worms quickly penetrate the skin and multiply

Fig. 38 The mayapple, a common woodland wildflower of the eastern United States, is grown commercially to provide materials for antitumor drugs.

within the human body. Millettia powder repels the worms; thus, it may be possible to incorporate it into an oil to be applied to the skin before entering infected water. The plant holds immense promise as a crop plant and has become part of the local economy through its medicinal properties. Since imported commercial snail killers cost up to $100 per kilogram, the savings will be significant.

Species of medicinal importance are by no means confined to rain forests or reefs; in fact, many of the most important ones come from very different environments. Aspirin is a simple derivative of salicylic acid, which was originally obtained from willow trees. Two new anticancer drugs have been isolated from the humble mayapple, a common herb in the eastern United States where, in the frigid depths of winter, rain forests and reefs seem a world away (see Figure 38). Some plant species from which classic drugs have been derived such as foxglove (digitoxin) and belladonna (atropine) come from cool-temperate climates; the poppy (codeine) is a Mediterranean species, and ephedra, from which ephedrine is derived, is a desert plant. A more recent discovery from a cooler climate is the powerful anticancer drug taxol, derived from the Pacific yew tree. It is intriguing that while the drug has been extracted from the inner bark, it may be a fungus inhabiting these tissues that produces the taxol.

Deep debate within the pharmaceutical industry about the future role of natural products has been prompted by the development of new methods of drug discovery that combine computers, robotics, and chemistry—the tools of what has come to be known as combinatorial chemistry. Imagine the diagram of a complicated molecule such as a promising new drug on a computer screen. With especially designed software the keyboard can alter the molecule or cause it to react with others without having to resort to the test tube to find out what happens. When subsequently a new molecular structure appears on screen, robotics takes over, synthesizing it and testing it against a range of disease organisms. Automation by robotics is now so rapid that thousands of potential new drug molecules can be tested every day in a single laboratory.

In the face of this technological wizardry, we may ask if there is any need for rain forests, coral reefs, or other wild places where natural medicines are found. For at least three reasons we think that the answer is yes. In the first place, as we have emphasized elsewhere in this book, the process of evolution by natural selection is the wild equivalent of combinatorial chemistry. Thus, if we look at an individual natural product derived from a plant, coral, or ant, we know that the development time has been millions of years rather than a few months; we know that various combinations and permutations have been tested over thousands of generations; and last but not least, because the species is here today, we know that the product works and is safe for its owner.

Second, organisms as diverse as plants, corals, and ants have evolved from profoundly different ancestors and have very different kinds of bodies. Furthermore, within each of these major groups very different species inhabit very different environments. For example, within the plants there are 100-meter-high tropical rain-forest trees, desert pines, seaweeds, temperate-zone tree ferns, and Arctic mosses. The evolutionary process has generated diversity on a scale that is hard to imagine and impossible to re-create in the laboratory. Thus, natural diversity still provides crucial leads for pharmaceutical innovation.

Third, the exploration of natural products can be greatly improved by applying evolutionary biology and ecology to the process. As we showed in Chapter 9, when we asked where the desired product may have evolved, the screening process that has been so inefficient in the past can be focused on the habitats or groups of species that for logical reasons should be sources of the desired product.

We suspect that one smart way to move ahead is to combine evolutionary and ecological methods with combinatorial chemistry. Screening of wild species that have been selected by evolutionary and ecological logic may provide the leads to or the inspiration for the molecules that chemists can then manipulate and test in vast numbers. Future clues are likely to come from a far greater array of organisms than in the past, but this fact may not be easy to appreciate. For example, it is easy to dismiss the bacteria and fungi as candidates for drug discovery because they have already been explored by modern industry. Yet, the reality is that these two groups contain so many species with such a vast array of life styles and capabilities that the industry is only in its infancy. Millions of species (or the bacterial and fungal equivalents of species) inhabit the world, and we need some form of logic to focus the screening process and to make it efficient and—dare we say it—cost effective. Add, too, other groups such as beetles and mites plus, of course, the plants (large, small, or single-celled) that contain huge numbers of species and equally varied life styles, and the promise of natural products becomes a lot more obvious.

Malaria

One of the most intriguing examples arises from an intimate knowledge of the diversity and biology of mosquitoes. While this field of study may seem both obscure and unpleasant, over the years it has revealed the vital intelligence that mosquitoes carry microscopic disease organisms in their bodies and that when they pierce the skin to take a blood meal, the microbes can be injected into the blood. In the case of malaria and many other mosquito-borne diseases, this bite is the start of severe illness during which the microbe multiplies prodigiously inside the body of the patient. The malarial parasite is a single-celled organism that spends time in the gut of some mosquito species, eventually producing infective spores that accumulate in the salivary gland. Thus, when the mosquito injects a little spurt of saliva into the tiny hole it has made in a human blood vessel, the spores gain access to the human body—a wonderful habitat for the parasite that is warm, safe, and full of food. The saliva is a vital part of the mosquito's operation in that it contains an anticoagulant that prevents the wound from clotting before the animal has completed its meal.

For years malaria has been fought with some success by spraying pesticides on the habitats where mosquitoes breed and by treating patients with antimalarial drugs. Unfortunately, both mosquitoes and parasites have developed resistance to these chemicals, and medical scientists are examining a range of new ways to attack the malarial parasite. One group has asked the kind of question that has become familiar in this book: Where might we expect to find the evolution of natural immunity to the malarial parasite?

The answer began to emerge as researchers realized they had simply overlooked the fact that malaria was also a disease of mosquitoes, and that parasites had been invading their intestines and clogging their salivary glands for millions of years. Had immunity evolved in mosquitoes? One obscure observation suddenly became pivotal to the entire investigation: there was diversity within a single mosquito species, because the bite of some individuals transmitted the disease but the bite of others did not. This discovery prompted other questions. Do the mosquitoes that fail to transmit the disease kill the parasites in their own bodies? In other words, could it be that some mosquitoes have already evolved resistance to the parasite? And if they have, could the mechanism be adapted for our use?

The answers are "yes" to the first two questions and "we don't know" to the third. Research has shown that the immune system of some strains of the mosquito *Anopheles gambiae* destroy malarial parasites in the gut, attacking them with specific chemicals. How this knowledge might be applied for human use is still uncertain, but identification of the chemicals that are induced by the presence of the parasite in the mosquito's blood is an essential first step. Following that, the application of sophisticated chemistry to the synthesis of these chemicals might lead to a drug suitable for clinical treatment. It may well be a long shot; yet the basic idea, although it may have seemed bizarre at first, has been taken very seriously and research teams in various parts of the world are pursuing it. If the concept proves out, then the solution may indeed have already evolved in the wild.

Other approaches to fly-borne parasitic diseases also take advantage of intimate knowledge of the natural history of the carriers. The tsetse fly, which transmits the trypanosome parasite that causes the human disease known as sleeping sickness, harbors bacteria in its gut.

They are important to the fly; if the bacteria are removed during laboratory experiments, the fly fails to reproduce. With the recent development of new molecular techniques, it has been possible to analyze the DNA of these bacteria with a couple of strategies in mind: either change the bacterial DNA so that they become a liability rather than a benefit, or load the bacteria with an antibiotic that will kill the parasite. In either case, the modified bacteria have to spread through the fly population. While that appears to be a serious obstacle, it is clear that the diseases that naturally attack the flies spread very rapidly, so the mechanism for dispersal of the doctored bacteria may already exist. This kind of research, both in the laboratory and in the field, faces many challenges. If it succeeds, however, it is potentially more environmentally friendly than the use of pesticides.

Leishmaniasis is another major tropical disease that is caused by a protozoan carried by a biting fly, in this case a sand fly. Experimental mice bitten by flies carrying the microbe soon contract the disease. Yet if they are first bitten by a fly apparently not carrying the disease, they tend not to become ill. This curious observation suggests that there might be something in sand-fly saliva that could help control the disease. We do not know if this is similar to the mosquito-malaria situation, wherein some flies have evolved resistance and others have not, but further research will sort it out. While sand-fly saliva might seem an odd biological or medical resource, it is of sufficient interest to be the focus of an intensive research program.

The idea that a cure may be found by studying the disease itself is, of course, a very old one. Here we ask where a cure, in this case resistance to the disease organisms, may have evolved naturally. We can go a step further, as many researchers have done, and ask if the disease-causing organism can be the source of a cure. One obvious answer is vaccination, because vaccines are made of living disease organisms that have been disabled in some way. When injected into the human body, the organisms in the vaccine are too weak to cause illness but they stimulate the immune system to make the antibodies that will repel a future attack. This science has recently been expanded by the study of deadly human-disease bacteria such as those that cause food poisoning, cholera, meningitis, and dysentery. All possess a special group of virulence genes that enable them to fight our immune systems and burrow their way into our organs and tissues, in turn causing the terrible symptoms of the diseases. However, laboratory experiments with the bacteria that cause food poisoning (*Salmonella*) have revealed the presence of a remarkable gene that appears to act like a master switch. When it is removed, the virulence genes do not work. The bacteria may invade, but no symptoms occur. Mice can be fed huge doses without experiencing any ill effects. Research into a new generation of vaccines made of bacteria with their master switches removed is racing ahead.

Invertebrates: The Good Guys

Leeches are probably the best-known invertebrate "good guys" in medicine, and still they have to overcome their image as blood-sucking worms that infest swamps and backwaters. Throughout recorded history from ancient Egypt and Rome up to and including nineteenth-century Europe, leeches have been widely used as a general cure-all to suck the "bad blood" out of the sick. Yet modern medicine has been deeply skeptical of their benefits. Anticoagulants or chemicals that safely prevent blood from clotting are in great demand, especially in Western societies where clots in blood vessels are a prime cause of illness and death. Leeches open a small wound in the skin through which the blood flows and they keep it open by secreting anticlotting agents in their saliva to aid digestion. The same chemicals prevent the blood meal from clotting in the leech's stomach. Thus, "leeches" are one answer to the question, Where might clinically useful anticoagulants evolve?

The pharmaceutical industry has taken leech anticoagulants very seriously for some years, as they have proved to have few side effects (Figure 39). Availability has emerged as a problem, though, for the leeches have been collected to extinction in some localities. Further, the most satisfactory species, the medicinal leech or *Hirudo medicinalis,* does not reproduce very rapidly and many of its favorite habitats have been drained for agricultural and urban development. In response, a company in the United Kingdom combed the leech biodi-

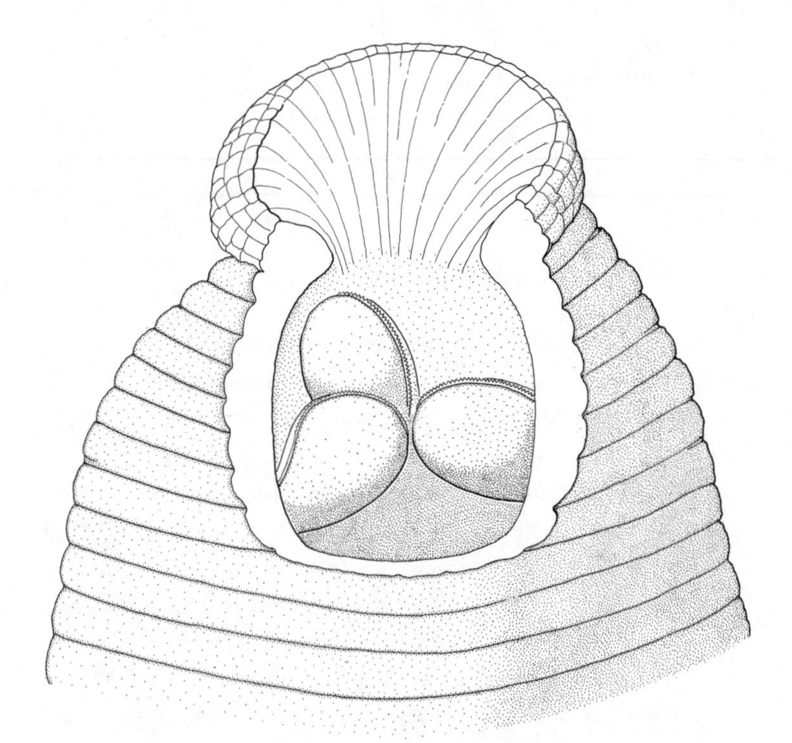

Fig. 39 The front end of a leech cut away to show the jaws. Although their appearance is far from delicate, they mostly go about their work unnoticed. Their painless approach is much appreciated by patients undergoing leech therapy.

versity of the world for a species that combines all the desired features, including the production of powerful anticoagulants, a high breeding rate, and secure populations in the wild. They found the giant Amazon leech in French Guiana and have extracted new anticoagulants from it. Inspired by their chemical structure, researchers have synthesized new generations of these valuable medicines, which can safely dissolve the blood clots that lead to strokes and heart attacks, two great killers of modern humankind.

The usefulness of leeches is not limited to their chemistry. In fact, the ancient belief in their utility has undergone a form of revival, as they are being used once again for their wholesome, old-fashioned ability to suck blood. They are employed postoperatively in reconstructive and transplant surgery, where bleeding into the wound causes serious problems. In the situation where a severed finger is being reattached, the microsurgeon has relatively little difficulty in finding and joining arteries because they are fairly robust pipes; the veins are more flimsy and easily missed. This means that blood flows into the joint better than it flows out, so that it becomes swollen. A large leech applied to the swelling will turn the wound from an ugly dark purple to a healthy pink. Further, the tiny puncture made by the animal contains a mild anesthetic so that the patient feels nothing, and an antibiotic so that the wound stays clean. Meanwhile, the leech, by this time distended to the size of a cigar, can be set aside to absorb its meal, and, when it is finished, the little sucker may be recycled to relieve another patient.

Lesser-known invertebrates valuable to medicine include caterpillars and fruit-fly maggots. As any gardener know, caterpillars sometimes come in plague numbers, quickly defoliating plants and rapidly putting on weight before pupation and metamorphosis. Researchers wondered if it might be inexpensive and efficient to get these eating machines to make medicinally useful proteins in addition to their own. The problem was to get the gene that coded for the useful protein into the caterpillar. Once again, knowing enough biology made it possible to identify a solution that had already evolved. This was a special kind of virus, a baculovirus, which lives by invading the caterpillar and persuading it to make virus proteins rather than caterpillar proteins. The end product is a caterpillar corpse that releases millions of new viruses available for infecting more caterpillars.

Molecular biologists have been able to insert into the baculovirus genes that code for medicinally useful proteins, among them the anticancer drug interleukin-2, insulin, and human growth factor. The virus is then let loose in a cage of caterpillars. Under the command of the modified virus, the caterpillars willingly manufacture the desired protein. The difficulty comes in pinpointing the precise moment when the caterpillar has made all the useful protein it can before dying from the infection. The addition to the virus of a gene for a fluorescent dye helps, as the color of the caterpillar changes in accordance with its protein makeup.

This technology is going through rapid development, and researchers are experimenting with a wide variety of caterpillars. Some laboratories are even turning to fruit-fly maggots, which do not mind crowded conditions and survive happily on simple diets. In addition, because fruit-fly genes are well known, the technology does not require a virus as an intermediary. The procedure is fascinating: the gene for the desired protein plus the gene that makes jellyfish fluoresce plus a gene that turns the fruit fly's eyes orange are injected into the embryo of a fruit fly destined to have white eyes. If the process works, the flies that have successfully adopted the useful protein gene will have orange eyes, and those making the largest amount of desired protein will fluoresce the brightest. These can then be bred together to produce races of superflies, with maggots churning out large amounts of the medicine.

In looking at the benefits of biodiversity for brain research, we move from caterpillars, fruit flies, and jellyfish to roundworms, sea butterflies, sea slugs, and back to our old friends the leeches. The human brain is large and extremely complicated, so it is very difficult to tease apart the mechanisms that run it. Thus, the question for researchers is, where can we find an animal with a well-defined nervous system that has such a limited number of nerve cells that we can keep track of them all? Brain research has focused on invertebrates because some have simple brains and nervous systems. The roundworm *Caenorhabditis elegans,* common in compost heaps, has about 19,000 genes and is composed of just 959 cells, about 300 of which make up the nervous system of the mature animal. Furthermore, its genetics are quite well understood. It is possible to alter one gene at a time and see what happens—what behavior is changed or what enzyme is no longer produced. Genes have been identified that prolong the life span, speed up swimming, or slow down feeding in the individuals in which they are active.

One use of this worm in brain research has been investigation of the ways in which brains gradually die. The hope is that understanding the process in the roundworm will shed light on what happens to us, particularly if we have a neurodegenerative disease that breaks down brain cells. About a quarter of the nerve cells with which the roundworm is born inevitably die within about three days, and the genes responsible for this outcome have been identified. These genes appear to be normal, and the animal apparently does not suffer from this "programmed cell death." Other genes, when they mutate, produce abnormal enzymes that kill specific nerve cells. Both varieties are of extreme interest in that they destroy nerve cells in the brain. Sea slugs, leeches, and other animals with simple nervous systems are being investigated with similar goals in mind. In spite of the huge differences between these animals and human beings, the invertebrates may serve as simple models that can be used to work out the mechanisms of much more complicated nervous systems.

Diseases of the nervous system have many causes. There may be a surplus of the chemicals that produce impulses from nerve cell to nerve cell (called neurotransmitters), or at other times there may be a shortage. Drugs that safely control the nervous system, or parts of it, are in constant demand, and lately medical researchers have looked to the natural world for inspiration. The concept of spiders, scorpions, or cone shell snails as likely sources may seem bizarre, but many of these animals produce venoms that kill by disrupting the nervous systems of their prey. Investigation of these venoms has revealed an extraordinary diversity of chemicals. The venoms, or artificial derivatives made in the laboratory, appear to have an extremely useful future in the treatment of Parkinson's disease, schizophrenia, depression, stroke, epilepsy, and severe pain. Once again, it is the sheer diversity of venoms that impresses researchers. There are tens of thousands of species of spiders and scorpions. There are about five hundred species of cone shell snails. While these may seem to be relatively small numbers, each species secretes up to two hundred different toxins!

The biology of venoms may reveal even more practical applications. Many spiders and predatory wasps prefer their meat to be fresh, which means that when they bite or sting their prey it is paralyzed rather than killed. Gruesome technique or not, they have evolved sophisticated venoms that subdue their meal without killing it. This feature may be useful to microsurgeons whose delicate operations require the short-term immobilization of muscle. It is possible to imagine a small surgical area being infused with a tiny quantity of wasp venom (or an anesthetic derived from it). It is immobilized while the surgeon performs a delicate procedure, yet movement returns promptly as the effects wear off.

One of the great risks of organ transplants is that the immune system tends to respond to the donated organ as though to a foreign invader. In a sense, this is an advantage; it is important to have a healthy immune system to repel unwanted incursions by microbes. However, in the case of transplants, there is no choice but to suppress the immune system long enough for the body to accept the new organ. Thus, chemicals that suppress the immune system safely (called immunosuppressants) are an extremely valuable commodity.

Where have they evolved? Answers that spring to mind are, in fungi and some parasites. Although this section is about invertebrates, we are going to make fungi honorary invertebrates for the moment because the basic evolutionary question is the same. Internal parasites that include a wide variety of flatworms and roundworms penetrate into the depths of their hosts, apparently without harm. How do they manage to invade intestines, lungs, kidneys, and brain without incurring the wrath of the immune system? Fungi have a similar life style, as their bodies consist of threadlike filaments that penetrate their food and form a network inside it. Digestive enzymes secreted from the filaments dissolve the surrounding food, forming a fluid that can be absorbed through the filament walls.

The food of fungi takes diverse forms. For many species it is dead material such as logs, leaves, or corpses and it is their activity that is partly responsible for the process we know as decay or rotting. However, many species invade living animals as varied as fish, frogs, and mammals. To do so, they too must avoid the activity of their victims' immune systems. The most famous immunosuppressant chemical is cyclosporin, which was isolated from a fungus. Further research into these invasive and parasitic organisms has revealed a variety of other immunosuppressant chemicals. Given the diversity of these organisms, their products or synthetic derivatives thereof are likely to be of medical importance in the future.

Medicines Selected by Animals

Our remote ancestors may have observed wild animals to discover plant species that could be used for food, others that should be avoided, and perhaps some that cured sickness. The behavior of some wild animals suggests knowledge of medicinal plants, and early humans may have learned by watching them. Certainly monkeys know how to select plants and other foods to maintain a diet that contains the major food groups and a variety of vitamins. The evidence that animals know how to select medicinal plants is intriguing, but difficult to interpret. Chimpanzees roll the leaves of a particular plant slowly and carefully around the insides of their mouths for up to half a minute before swallowing them. The foliage belongs to the hemorrhage plant, so called because its juices stop bleeding, but it also contains antibiotics effective against a variety of bacteria, fungi, and parasitic worms. Sick chimpanzees also search for and eat the bitterleaf plant, another species that contains antibacterial and antiparasite chemicals. Bears, deer, elk, and various carnivores such as coyotes, foxes, and cougars have all been observed to seek out and consume specific plant species, and these observations have been interpreted as the use of medicinal plants by wild animals. If this interpretation is correct, it may be useful to observe sick animals and analyze the plants they select. But even if a sick animal, say a gorilla, obviously uses a plant that is not part of its normal diet and gets better, it is difficult to know exactly what the gorilla was suffering from in the first place. This unavoidable lack of communication will make it difficult to match the medicine with the disease. While field naturalists have many anecdotes that can be interpreted as self-medication by animals, the field is very new and there is as yet little reliable information. Whether or not watching animals will prove to be a useful method of discovering new medicinal plants remains to be seen. Still, there is little harm in collecting materials that appear to be used medicinally and having them analyzed for their pharmaceutical properties.

Viruses

Viruses are a familiar problem indeed, for they cause major and widespread disease in people, domestic animals, and crops. Yet they may also provide solutions in the form of cures for some diseases, and these may be found by understanding their ecology. Viruses are the submicroscopic equivalent of predators that live by attacking and destroying bacteria and many other kinds of cells. Perhaps, then, viruses can be used to attack the microbes that cause disease or the animals such as mosquitoes that spread them. In experiments where mice have been infected with a kind of bacterium that would normally cause their death, but were then injected with a virus that specifically kills that bacterium, the mice recovered. A valuable characteristic of many viruses is that while they are effective predators, they are extremely limited in the range of bacteria or cells they can attack. Thus, the deadly disease used on the trial mice was a strain of *E. coli;* it was killed without harming the beneficial strains of *E. coli* that inhabit their digestive systems. These kinds of experiments will continue in medical laboratories, as viruses that kill disease-causing bacteria offer an alternative to conventional antibiotic treatment. There is a fine irony here: the idea of administering viruses to patients was first introduced eighty years ago, but the research ceased with the introduction of antibiotics. Now that antibiotics are losing their effectiveness, we are turning once again to these minute microbe killers.

THIRTEEN

Savings and Loans

The biodiversity of Earth is our biological wealth, our biological capital. The savings are every gene, every population, every species, and every natural community that inhabits the oceans, the land, and the air. Whether we believe that God put them there or that they evolved from earlier creatures, the stark truth remains that they are the only ones we have—there are no life forms anywhere else. As yet there is no evidence whatsoever that one day humans will be able to fly to Mars or some other remote planet to stock up on tree species, order giant panda replacements, or obtain refills of extinct phyla. Nor is there any hope, based on the current status of biotechnology, that we will be able to create organisms through the miracles of science. Biodiversity is, as far as anyone knows, totally irreplaceable. It would be marvelous to be proved wrong, but we're not holding our breath.

This book has considered a wide variety of loans that humanity has taken out from the biological capital of Earth. Their extent is staggering: entire natural communities, each made up of hundreds of thousands of species, that—worldwide—maintain the fertility of soils, dispose of wastes, preserve the quality of the water and the air, and combat the pests that attack crops and forests. The loans remain possible as long as the capital remains intact. Diminish the capital and the loans become chancy, that is, the ecosystem services decline or cease.

Before we become too carried away with financial analogies, we must emphasize that while human financial laws and institutions are changed on a regular basis, natural laws governing the evolution and survival of species are immutable. This means that the savings and loans concept we use here is much less forgiving than the human transactions on which the analogy is based. When species go extinct, there is no counterpart to refinancing or debt forgiveness; the species are gone, along with the services they provide.

Although most people probably have not thought in these terms before, ecosystem services are on loan from Nature. Every industry—agriculture, forestry, fisheries, tourism, manufacturing, banking, and insurance—is a wholly owned subsidiary of one or more natural communities. The same is true of the glamour occupations such as professional sports, filmmaking, the stock market, and major institutions such as the church and the military. In addition to ecosystem services, biodiversity has loaned us a vast range of commercially valuable species. We take most for granted, with little thought for their origins. They include our farm animals and crops, together with their relatives in the wild from which we harvest the genes that maintain agricultural productivity in the face of changing markets, pests, and climates. However, in this book we have shown that the number and variety of species that have been and will be useful to humanity are far greater than previously imagined.

In *Wild Solutions* we have discussed the benefits of hundreds of individual species of viruses, bacteria, fungi, roundworms, earthworms, single-celled animals, corals, rotifers, spiders, tardigrades, crustaceans, mollusks, insects, vertebrates, lichens, algae, and many species of flowering plants. Not all of the ninety-six phyla are known to be "useful" in the sense that they contribute directly to human welfare through ecosystem services or resources, but many more are involved than most people think. Further, numerous examples of industrial innovation show repeatedly that we can never predict what species or populations are going to be useful, or even desperately important, in the future. Our spotlight here has often been directed at the most unlikely, obscure, or bizarre species.

It has become fashionable in some circles to assert that humanity cannot possibly need all the species on Earth. This may turn out to be true one day, but which ones are redundant? We know that ecosystems evolved for billions of years before humans inhabited the planet. They did not need us, and they will carry on with or without us. Yet when the context includes people, the situation changes. People do need ecosystem services, each of which involves an unknown number of species. The key question is, how many species are needed to maintain the services that keep us alive? It seems that some extinctions do not much affect those services; for example, millions of people live in Europe where nearly all the large predators have been eliminated. The tragedy of loss in these cases is no less because it is a moral or aesthetic issue, but ecosystem services to date show little sign of detrimental effects.

We get glimpses into the question of how many species we need from accidental or intentional abuse of ecosystem services, as when pesticides are overused or too many wastes are dumped into rivers or the sea. Species then are killed in large numbers and the services often fail. It would be highly advantageous if some of these accidents or disasters were treated as experiments—so that at the very least we could determine which species were present, and in what numbers, before and after the event. Thus, we might know which species are missing when the service declines or breaks down. Sometimes the answers seem obvious, as when floods and mudslides or salination of soils follows the destruction of vegetation, especially of forests and woodlands. Still, what the critical number of species needed may be, or which species may turn out to be the most important, and at what time and place, we simply do not know.

We think it is wiser to turn the whole concept of redundancy around. Thus, in the same way that "redundant" systems are engineered into modern passenger jets as insurance against electrical, engine, or structural failure, "redundant" species provide insurance against failure in the natural systems on which we depend. We are suggesting that *any* species could be important, sooner or later. Who knows, for example, what tree species will emerge as our main source of timber as the climate continues to change? Who knows which beetle or wasp will emerge to control pests when pesticides and biotechnological solutions fail? What species of marine animals will form the aquaculture of the future, or which bizarre species will provide an essential industrial or medical chemical?

Right now the human race is recklessly squandering its savings. All over the world forests, soils, and fisheries are in decline. The oceans, our freshwater, soils, and air are polluted in various ways. We are even changing the climate for the worse (or to put it another way, weather that we always considered a problem is likely to become a nightmare). What will our children and grandchildren do if the capital is used up?

The answer is to make sure we do not lose biodiversity, and that means conservation. We have seen that we all depend on the millions of genes, populations, species, and natural communities that we call biodiversity. Conservation is not just for environmentalists, it is everyone's business. Who can afford to ignore the natural processes that keep us alive?

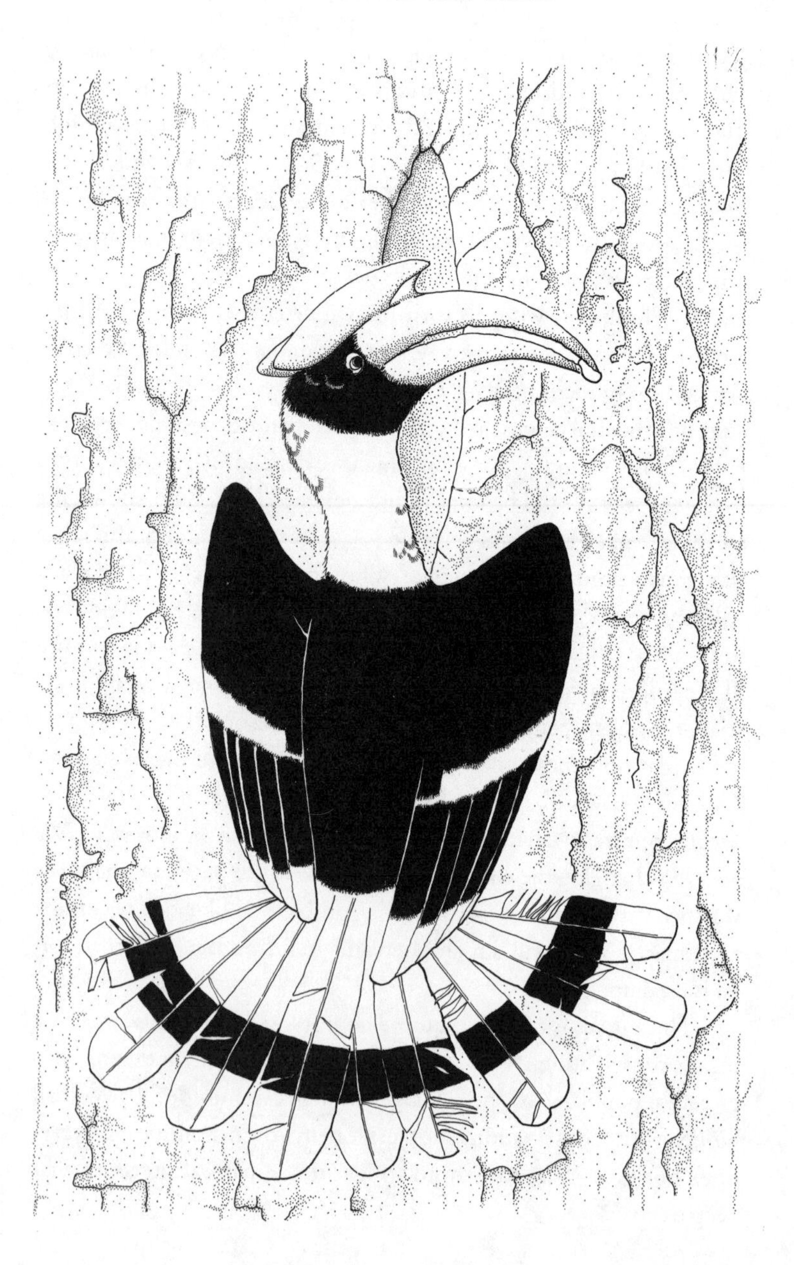

Two major ingredients of conservation are imagination and action, as dramatically demonstrated by Pilai Poonswad in Thailand. She was faced with gangs of poachers taking hornbills—huge, beautiful, rare birds of the rain forest (Figure 40). Unscrupulous collectors paid large sums for eggs, chicks, or adults and the poachers defended their booty with violence when necessary. Pilai went to the villages where the poachers lived and showed them that the birds were worth far more alive than dead. Today she has a growing band of *ex*-poachers earning far more by taking tourists into the forest to see live hornbills than they ever made by stealing them. A superb example! The birds and their rain-forest habitats remain intact, the local people profit from the presence of the undisturbed forest, and visitors from all over the world learn a little about biodiversity. The venture is funded by tourists and by city dwellers in Bangkok who sponsor individual nests.

We all know the three *O*s that cause the decline of biodiversity and the extinction of species: Overpopulation, Overdevelopment, and Overconsumption. The global human enterprise uses unsustainable technologies in agriculture, forestry, fisheries, and industry and produces massive urban developments generating worldwide pollution. Although based on biodiversity, all these activities contribute daily to its decline. The developed countries have the wealth and power to influence economic activity on a vast scale, yet their actions are subversive in the very real sense that we are steadily eroding our own capital.

This book shows why so many people have accepted that our most important capital is natural capital: biodiversity. To recapitulate

Fig. 40 A hornbill from Thailand at the entrance to its nest in a tree trunk. Poachers once stole these magnificent protected birds and sold them or their feathers to unscrupulous collectors for a meager income. Today they take tourists into the forest to see the birds in their natural habitat and make a decent living from their fees.

one of our early themes, *this* is the capital of the real world. The currencies based on financial capital derive from the life-support systems and products generated by biodiversity. We know full well that humanity has chosen to operate in a world of financial capital. The way to proceed, therefore, is first to recognize the true worth of the two kinds of capital and then to organize the human economy to preserve both. The value of financial capital is accepted by most people; the value of natural capital is only starting to be recognized. Our hope is that the *concept* of natural capital is accepted before its *value* is destroyed. That is why we wrote this book.

Afterword

"It was the best of times; it was the worst of times"—the opening sentence of Charles Dickens's *A Tale of Two Cities,* published in 1859, resonates ominously today. Two global revolutions are taking the planet in opposite directions. The first is the accelerating rate of the discovery of new species. The second is the continuing loss of species through extinction; thousands are expected to disappear this century. The terrible irony is that as new biological technologies reveal thousands of new species each year, the dominant social, economic, and political forces of the world continue to destroy the environments in which they live. To draw an analogy from economics, we are discovering resources and creating jobs—and destroying them.

The first revolution, the increasing discovery of new species, has astonished everyone, including biologists, and is largely the product of new technologies for the exploration of species that live in the oceans. Discovery of those species in the 65 percent of Earth's surface that is covered by seawater has been slow until now, but suddenly marine scientists are finding new species living everywhere in the open, apparently lifeless ocean—at great depths, within bottom mud and sediment, and, of course, in coral reefs. Most are microscopic—bacteria, viruses, and fungi, along with many kinds of single-celled animals and plants. The numbers of individual organisms can be staggering; for example, an average milliliter of seawater—that's a volume the size of a sugar cube—contains a million bacteria and ten million viruses. The actual number of species will take a long time to determine. In fact, the very concept of species may be impossible to apply, because microbes exchange genetic material so regularly and so widely that genes that might define species are constantly on the move. Because microbial species are so tricky to identify, research has focused more on what they do. One example reveals the potential for wild solutions: in Chapter 11, "Blueprints and Inspiration from the Wild," we talked about the phototrophic and photosynthetic machinery of bacteria and plants and how engineers were trying to mimic these highly evolved mechanisms to find inexpensive, large-scale, and environmentally friendly ways of using solar energy. While significant progress has been made in the past three years, the recent discovery of new marine bacteria and algae is likely to provide a boost to the research effort. These organisms possess previously unknown mechanisms for converting solar energy to chemical energy, and the chemical mechanisms may serve as blueprints for ongoing research into the cheap and efficient conversion of solar power for human use.

These exciting marine microbial discoveries also create significant new insights into the "natural internet" (Chapter 4). It is not easy to see how such tiny organisms can exert global effects, but if you were piloting a microscopic ocean submersible, you would see nature, perhaps not "red in tooth and claw," but with all the drama you expect when wild organisms interact. One of the most spectacular interactions is the war between viruses and algae. Viruses, the predators of the microbial world, attack all kinds of cells, including marine algae. Algal populations often undergo explosive growth or "blooms," fueled by natural nutrients or by escaped manufactured ones such as farm fertilizers. These present a feast for the viruses, which attack and kill them in enormous numbers. In some cases, when the slaughter is massive, the billions of decaying algal corpses release gases into the atmosphere in such vast amounts as to stimulate cloud formation. The clouds reflect sunlight, and this, in turn, cools the climate—a process that illustrates that even microbes, in sufficient numbers, affect planetary processes.

Although we have focused on microbes so far, new larger marine animals and plants are being described at the rate of about 1,700 per year. Perhaps the most astonishing is *Balaenoptera omurai,* a twelve-meter whale from the Sea of Japan and the eastern Indian Ocean. The description of new fish and invertebrate species, especially roundworms (nematodes), which we talked about in Chapter 2, is now commonplace. The numbers are awesome: it is estimated that there are 210,000 marine species known to science, but ten times that many probably exist. There may be a million species of marine nematodes alone. The current total of 15,300 known marine fish seems likely to grow past 20,000. As new technology enables the exploration of Earth's largest habitat, the seabed, including its ultradeep habitats and volcanic vents ("black smokers"), the biodiversity of this vast area and the implications for wild solutions are gradually being revealed. A recent exciting wrinkle is the exploration of undersea mountain chains (seamounts) that, like their terrestrial equivalents, are especially rich in endemics—that is, species restricted to a particular mountain. There are an estimated 30,000 seamounts worldwide, promising a wealth of new marine "alpine" species.

There have also been many spectacular discoveries on land, not least a new Order of insects from Namibia, Tanzania, and South Africa. The rank of Order is assigned to such major groups as the dragonflies, grasshoppers, termites, cockroaches, beetles, bees, and butterflies and moths, each of which is so distinctive as to be easily recognized by most people. How could an Order have been missed during the recent centuries, while entomologists have scoured every corner of the globe for their collections? To be fair, the new Order—the Mantophasmatodea (see Figure 41)—does look a little like a cross between its major close relatives, the stick insects (Phasmatodea) and the rock

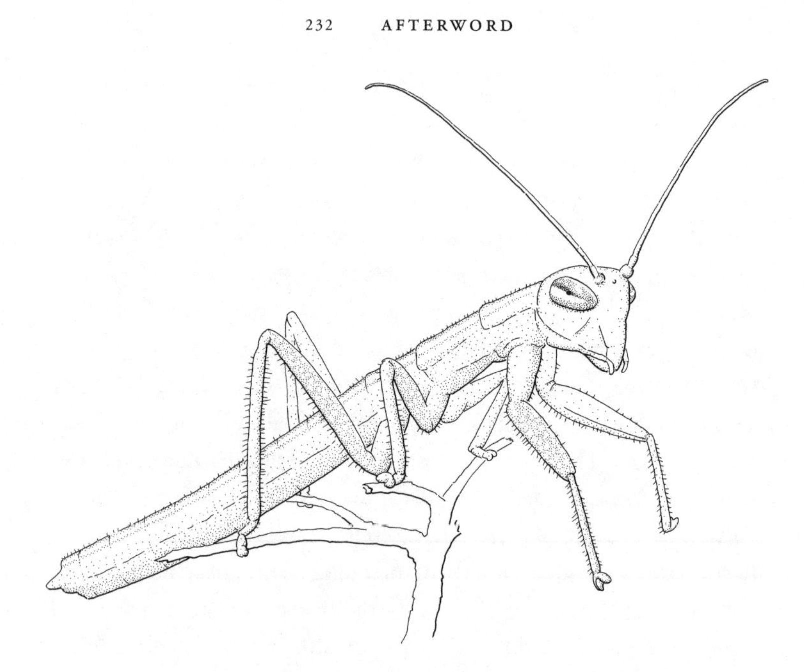

Fig. 41 The discovery of the century: a mantophasmatodean from South Africa, a member of a brand new order of insects.

crawlers (Grylloblattodea), but now that scientists have become alerted to their presence, it turns out they are locally common in some areas of southern and eastern Africa, and a trawl through museum collections has revealed many (mislabled) specimens.

Frogs make up another well-studied and much-loved group of animals. As most frog families were named by the mid-nineteenth century, it was sensational news when a new Family was described in 2003. This recent addition hails from the Western Ghats of southern India: plump, stumpy legged, and short-snouted, it reminds us that even among the best-known groups, important discoveries remain. Our final example in this section may seem the most unlikely of all. The butterflies of the United Kingdom are among the most studied of all animals. For several centuries both amateur and professional butterfly enthusiasts have wielded their nets throughout the Kingdom, generating huge collections and a literature so large and excruciatingly detailed that presumably nothing should have escaped scrutiny. And yet during 2002 a species never before known from Britain was discovered there, the wood white *(Leptidia reali).* At first it was thought that the butterfly might come from adjacent habitats on the Continent across the English Channel. But it does not live there; the nearest populations occur in southern Europe, so its sudden discovery has proved both delightful and mysterious.

The second revolution, the disheartening downside of all this discovery, is the acceleration of the loss of biodiversity. This is not just the well-advertised extinction of species but the equally important loss of populations. The extinction of populations almost always precedes the extinction of species, and populations often differ from one another in ways that are important from the perspective of wild solutions; one population of a species, for example, might have individuals with a concentration of defensive chemicals that are not present in the species as a whole. Thus population losses not only signal eventual species extinctions but also represent a depreciation of biological capital. But equally important, or maybe more so, is that the existence of a large number of populations is crucial for the delivery of ecosystem services. A thought experiment readily illustrates this principle: Suppose all Earth's species were each miraculously reduced to single, minimum-sized populations that would not go extinct. This would guarantee that our planet's species diversity would be retained, but humanity would soon disappear for lack of food (since, for example, a minimum viable population of each kind of cereal could not support a single family). Remember that the existence somewhere of predators of crazy ants, cockroaches, and katydids was no help to the inhabitants of Biosphere 2, who needed the predators' pest-control service but lacked the necessary populations of insect eaters.

Loss of populations is most obvious to us at the moment in the decline of stocks of wild fish worldwide, as population after population is harvested to economic extinction—such a low population size that it no longer pays to fish them. Providing food to us from the sea has long been one of the most crucial of ecosystem services, and as fishing operations create havoc on the seabed, it appears much of that service will be lost, and with it many possibilities of wild solutions from marine communities.

Since *Wild Solutions* was written, signs of climate change have been increasingly apparent, and the scientific community is documenting the impacts of that change on biodiversity. The threat from climate change alone is extremely serious, but of course other negative trends—the burning, clearing, and fragmentation of tropical forests, for example, or the paving over of areas for development, the overuse of groundwater, and overfishing—are also major sources of biodiversity destruction. These sources of depreciation of our biological capital are, of course, associated with the "three O's." Barring catastrophe, overpopulation will continue to increase substantially. While population growth has slowed, it seems likely that at least two billion more people will be stressing human life support systems well before the end of this century. And there is no end in sight for escalating overconsumption and overdevelopment, especially among the rich. One need only consider the nearly doubled size of the average American home that has accompanied the more than doubled U.S. population since World War II. The increase in home size alone means more forests cut down to provide timber; more mineral development to provide for fixtures, appliances, and power to light, heat, and cool; and more wild populations exterminated in the process. Overconsumption seems to be a problem even more intractable than overpopulation, one closing off avenues to wild solutions from every side.

The scientific community has warned us in no uncertain terms about the consequences of our behavior. More than 1,500 of the most distinguished scientists, including more than half of the living Nobel laureates in science, signed a declaration in 1993 that proclaimed: "Human beings and the natural world are on a collision course. Human activities inflict harsh and often irreversible damage on the environment and on critical resources. If not checked, many of our current practices put at serious risk the future that we wish for human society and the plant and animal kingdoms, and may so alter the living world that it will be unable to sustain life in the manner that we know. Fundamental changes are urgent if we are to avoid the collision our present course will bring about." In the same year, fifty-eight of the world's science academies, including the U.S. National Academy of Sciences, the Royal Society, the Australian Academy of Science, the Chinese Academy of Sciences, the Indian National Science Academy, and the Third World Academy of Sciences, issued a similar statement: "The magnitude of the threat . . . is linked to human population size and resource use per person. Resource use, waste production, and environmental degradation are accelerated by population growth. They are further exacerbated by consumption habits. . . . With current technologies, present levels of consumption by the developed world are likely to lead to serious negative consequences for all countries. . . . As human numbers further increase, the potential for irreversible changes of far-reaching magnitude also increases."

So the scientific consensus was clear more than a decade ago, and yet humanity continues to destroy its precious biological capital. There are many reasons for this, such as the difficulty that human nervous systems seem to have in perceiving threats that accumulate gradually and affect us mostly over the long term. A more pernicious reason is the maldistribution and misuse of power. While some governments are struggling to generate laws and policies that protect the environment, others are actively destroying them. The abuse of political and economic power to suppress laws, policies, and public movements that favor the conservation of biodiversity is not confined to any one government and is also evident in the mass media and among big corporations. This pervasive abuse, often coupled with ignorance, erodes humanity's natural capital and, in the process, stifles the possibilities of developing many more wild solutions that could save the lives of billions of people and make those lives much safer and richer.

Progress in implementing wild solutions has been widely documented in the past three years. Professors Lissy Coley and Tom Kursar have pioneered wild solution methods for finding new pharmaceuticals in rain forests. Normally when pharmaceutical companies go rain-forest bioprospecting—that is, looking for plants that might contain new drugs—they collect as many as possible on the basis that the more plant material examined, the more likely something useful will be found. Or, in their language, the more screening, the more "hits." Leaves may be harvested at random so that many are large and mature. Usually they are then dried and ground into powder for laboratory testing. Coley and Kursar questioned these methods and showed that a little ecological knowledge applied to the collection phase led to major improvements. Their thinking, based on ecological theory, went like this: as there are more species in tropical rain forests than in any other land ecosystem, there must be more herbivores there than anywhere else. Or, to turn this hypothesis around, the need for defenses against herbivores must be greatest at these latitudes. And the development of defenses is particularly important for plants, because they are unable to run away or hide.

Plants commonly defend themselves by making their tissues as unpalatable as possible, so as leaves get older, they are increasingly loaded up with such tough, unpalatable chemicals as cellulose, lignin, and tannin, which are hard to bite or chew—or digest. These chemicals render leaves nutritionally uneconomical to eat and, in bulk, provide the physical strength that holds them in position. This is a perfect defense for old leaves but completely inappropriate for young expanding leaves with cells that must remain flexible as long as they are growing to final size. Instead, so the theory goes, young leaves should secrete a very different set of chemicals that actively repel and inhibit biting and chewing or even poison the attacker (see Figure 42). Coley and Kursar's research in the rain forests of Panama has confirmed this conjecture; they found that growing leaves contain a far richer variety of pharmacologically important compounds than older ones. This basic discovery meant that future collections can focus on young leaves and will save the time and effort previously wasted on older ones. They also abandoned the ground-up powder, which often lay around the lab for years, and extracted the chemicals from the young leaves while still fresh. The first result of applying these more cost-effective methods has been a series of promising leads for drugs active against several kinds of cancer and three tropical killers—leishmaniasis, malaria, and Chagas' disease. There have been two other major advantages: First, this ecologically driven bioprospecting is carried out in developing countries where the employment and commercial knowledge it generates stimulates local economies. Second, in the researchers' own words: "Drug discovery is a non-destructive use of biodiversity that creates incentives to conserve wildlands." This is especially true since not only will new techniques allow the discovery of more and more novel chemicals but the evolutionary "arms race" between plants and herbivores ensures that nature will always be inventing new drugs.

The Lotus Effect

Reading about one of the most dramatic wild solutions in recent years is likely to make you say: Why didn't I think of that? Most of us have watched rain falling on plants—during a downpour in the garden or while walking in the woods. Some of us may even have wondered why the plants don't appear to get wet. After all, animals clearly do—witness the bedraggled feathers of birds after a storm or the sodden fur of the family dog after running through wet grass. A few of us may have

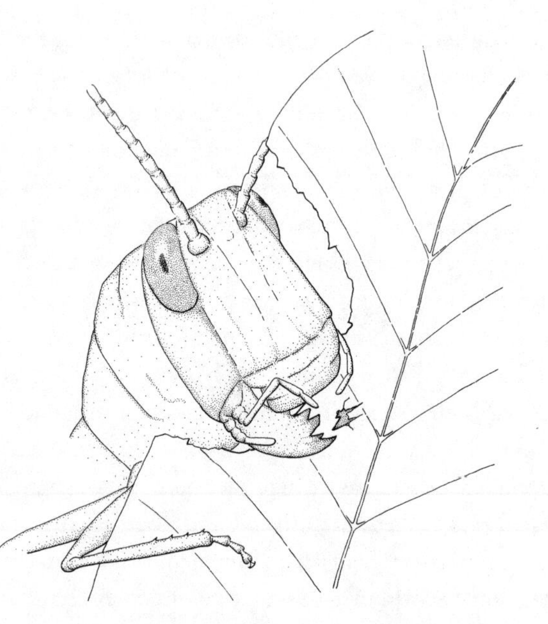

Fig. 42 Insect herbivores consume massive amounts of vegetation; in fact, they are the main enemies of plants. Here we see some of the apparatus insects use to bite and chew leaves.

even noticed that rain landing on leaves often forms "beads" and runs off onto the ground. Some may even have wondered how that works. Professor Wilhelm Barthlott went a lot further and examined the surface of the leaves of hundreds of plant species to find out how they remained both dry and clean. He eventually settled on studying the lotus, a tropical plant species with a leaf like a green umbrella that never seems to get wet or dirty. Under the microscope, he found that the surface of each leaf is covered by minute wax crystals or nanoparticles that gather water into tiny droplets that then roll away, taking any dirt with them, rather like snowballs. Here was the answer to the question: Where would you expect self-cleaning surfaces to have evolved? A European patent for this natural self-cleaning surface gave rise to the trademark "Lotus Effect," and research and development by a group of companies has yielded another two hundred patents and a series of self-cleaning products, the first a paint for external use on buildings, and others for textiles, paper, and leather. The leaf-surface blueprint and the inspiration from the lotus plant is giving rise to still more imaginative applications, such as self-cleaning surfaces for pipelines, glass, and maybe even cars.

Bacteria and the Business of Bugs

We have already addressed many aspects of the immense resources harbored by the vast and largely undescribed diversity of microbes. But this field of wild solutions has progressed so rapidly, and the industries involved are growing so explosively, that we should say a little more about it. The products generated by microbes of one sort or another are now worth many billions of dollars a year—nobody knows exactly how many. They include antibiotics, enzyme inhibitors, essential amino acids, immunosuppressants, anticancer agents, vitamins, biopesticides, and a vast range of highly specialized industrial and scientific chemicals. Most commonly, a gene that codes for the manufacture of the desired chemical is engineered into an appropriate bacterium, which is then cultured in huge quantities. Each bacterial or fungal cell is a minute factory, producing tiny quantities, but on an industrial scale millions of liters or kilograms can be produced, most of great purity. The great advantage of using these highly evolved mini-factories is that nature has long since worked out the complex series of steps required to make the final product. Although modern industrial chemistry is very sophisticated, it is still often cheaper and more efficient to hand over the manufacturing process to a bacterium or fungus rather than attempt to mimic it in the laboratory or factory. In many cases, this process would be otherwise too expensive and in some cases, the processes are so complex and subtle that chemical engineers have been unable to work out how nature does it.

The new species of microbes being discovered daily appear to have great potential as wild solutions. Take an unnamed species found near a hydrothermal vent in the Pacific; this hardy soul survived for two hours at 130°C, 30° above the boiling point of water. In contrast, entirely new microbial communities have been discovered at great depths in the sediments beneath Lake Vostok (which is, in turn, beneath the Antarctic ice) and in the layers of sediment and rock beneath the Pacific Ocean. Many of these are adapted to great pressures and very low temperatures. What incredible enzymes or products are yet to be found in these largely unknown cellular factories? You will note that many of the sites where new microbes are being found are extreme environments, and it is not surprising that others are being discovered in manmade extreme environments, like highly polluted areas. Mine wastes and abandoned factories continue to be the source of microbes capable of detoxifying poisonous pollutants, including the most infamous, worldwide dioxin. Finally, teams of scientists in several countries are exploring microbial ecology with a view to understanding how diseases might be cured by such microbial weaponry as viruses or predatory bacteria and fungi. While this kind of scenario generally gets negative press by association with biological warfare, a great deal of research is being carried out by respectable medical scientists seeking new ways to set one microbe upon another in the human body, and by agricultural scientists deploying predatory microbes into the tissues of diseased crops and farm animals where the viruses seek out their bacterial prey.

Molecular Biomimetics

"Molecular biomimetics" is a mouthful that refers to the application of wild solutions at the molecular level. It is part of the new and booming nanotechnology industry (nano = one billionth) in which machines are scaled down to such incredibly minute dimensions that individual molecules are the components. While nanotechnology has been largely developed through advances in the physical sciences, it should come as no surprise that it has turned to the small end of biology for inspiration and blueprints. After all, life itself began through the assembly of molecules into efficient, self-replicating machines called cells. Miniaturization has been the standard, and, as we saw in the previous section, bacterial cells have been employed as factories by a variety of industries to manufacture many different products.

Nanomachines are made of nanomaterials, and making nanomaterials is itself a major challenge that often requires three stages. The first is manipulating bacterial DNA to code for the exact protein required to function as the base layer. Second, a detailed knowledge of the surface of that protein is required so that other layers, either organic or inorganic, can be attached. Finally—and this may appear almost magical—because a portion of the nanomaterial is biological, nanotechnologists aim to select molecules that, as in a living organism, self-assemble and so continue generating the desired material in a highly ordered way. The result is often a hybrid nanomaterial, part organic (protein, say) and part inorganic (gold or titanium, for example). One such material has been developed to provide a base layer with a surface that holds and exposes diagnostic DNA molecules in medical samples to detect such dangerous bacteria as golden staph *(Staphylococcus aureus).* A molecule is incorporated that changes color when the bacterium is present. Diagnosis is rapid because the nanomaterial responds to minute quantities and does not require amplification of the DNA from the disease bacteria to create the larger amounts currently required for diagnosis. The test is not only rapid but ultrasensitive, making it possible to tell whether the bacterial sample under investigation is resistant to antibiotics.

The search for wild blueprints has taken nanotechnologists to many denizens of the natural world—to squids and algae, for example. Squids are well known for the spectacular ability to change color, either for camouflage or to signal aggression or danger. Recent research has shown that the Hawaiian bobtail squid manufactures a highly specialized nanomaterial that facilitates such rapid color changes. Such unique materials, as well as their protein components, recently named reflectins, will be useful blueprints for biosensors; and, being an adaptation to light processing, may be useful in the development of artificial photosynthesis. The minute, single-celled algae known as diatoms possess skeletons of silica, an element which is the principle component of glass. There are approximately ten thousand living species, many of them of unrivalled beauty, and the skeleton of each species is characteristic, with its own shape, size, and complexity. For the nanotechnologist their main interest lies in the ability to manufacture perfect three-dimensional silica skeletons with exquisite precision and in large numbers. Another feature is of still greater value: they can achieve this feat at the normal temperature and pressure of their ocean environment and without any polluting additives. Contrast the industrial processes of glassmaking and transforming glass into commercial products; these generally require enormous temperatures, sometimes altered pressures, and often the addition of extremely toxic chemicals. How diatoms manufacture their "glass" skeletons without any industrial paraphernalia is of profound interest as engineers envisage the manufacture of such nanosized machine parts as the cogs for minute gears or filters with precise pore sizes. The holy grail of this branch of the industry is the discovery of the diatom genes that achieve this astounding microengineering, heralding the possibility that single-celled algae may one day be used as the factories for the manufacture of nanosized machine parts to order.

How does a diatom make an inorganic (nonliving) skeletal structure using the organic, physiological processes in its body? We can ask similar questions about a wide variety of more familiar organisms:

How do they make bones, teeth, shells, spines, exoskeletons, or lenses, which are either entirely inorganic or have inorganic components? Many kinds of minerals are used, including calcium, silica, iron, magnesium, and barium; these occur as a variety of chemical compounds, including oxides, hydroxides, phosphates, and carbonates. Why have they evolved the way they have; what specific advantages did they bring to their hosts? But above all: how do they do it? These questions are the focus of an area of scientific research called biomineralization. We introduced this field in Chapter 11, and it has grown rapidly in the past few years partly because of the growing recognition of the idea of wild solutions among engineers and partly because of the advances in molecular biology that have fostered the rapid expansion of nanotechnology. As J. D. Birchall, one of the founders of the science, said: "Biology does not waste energy manipulating materials and structures that have no function and it eliminates those that do not function adequately and economically. It is well, then, to look for fresh insights to biology and the wisdom encapsulated in the materials it uses."

Biomineralization is the study of the chemical and physical processes by which living organisms make inorganic structures. It is by necessity a reductionist science whose practitioners seek to understand these processes at the level of individual atoms and molecules. Once the processes are understood, however, the applications go far beyond nanotechnology to the manufacture of sensory and optical instruments, advanced ceramic materials, and a wide variety of new organic-inorganic composite materials for construction and engineering. Recent research based on such crystalline structures as the spines and protective plates of starfish has generated new insights for the manufacture of the highly specialized crystals required for many electronic devices. In a very different direction, telecommunication engineers applied the idea of the wild solution to ask the question: where might you find the inspiration to improve modern telecommunication fibers? One answer is among deep-sea glass sponges, especially the beautiful animal known as the Venus flower basket *(Euplectella sp.)*. It is a long, thin cage constructed from a lattice of fused silica spicules or needles, each with a structure that strongly resembles commercial telecommunication fibers, but is far more resistant to fracture. In addition, as we have just seen with the silica diatom skeletons, *Euplectella* manufactures at normal temperatures and not the extreme heat required by the industrial process, a feature of immense commercial and technological importance. The engineers involved in this research are studying the natural processes in great detail, for if they can mimic those processes, they will be able to avoid the problems and expense of high-temperature manufacture and to exploit the chemicals, selected by the sponges over millions of years, that generate the superior properties of the fibers.

Restoration and Biomonitoring

The reconstruction of damaged ecosystems is one of the most exciting and rewarding biodiversity-based industries. While the goal of being able to entirely reassemble the species that once occupied a piece of the landscape may remain forever beyond the ability of contemporary science, we are increasingly able to select from the local biological diversity the species that work best together to create a lasting community. The focus of this selection is mostly on plant species, employed mainly to stabilize whatever soil there may be. But Professor Steven Handel advanced this research by selecting plant species that attacted birds. His restoration sites were the massive New York City landfills on Staten Island. His aim was to provide food for the berry-eating birds so that they would visit the restoration areas. Handel and his team reasoned that these birds would not depend upon his sites alone but would feed in far-flung plant communities as well. Thus viable seeds would already be in the birds' guts when they came to the restoration areas. This turned out to be the case, and as wild birds came to his carefully selected shrub communities to continue feeding, they left behind in their droppings a wide variety of seeds, which germinated the following spring. The combined efforts of restoration scientists and wild birds generated communities of plants well adapted to the local soils and climate. Little did the city realize how much it depended upon its feathered friends.

There has also been a growing recognition that the entire process of revegetation can be speeded up and rendered more sustainable by carefully selecting the appropriate soil microbes. (Chapter 6 has a relevant section on soil fungi that are symbiotic with plant roots.) One of the major problems of establishing plants in sites undergoing restoration is that they suffer heat stress because there is little or no other vegetation, and thus bare soil is exposed to the sun. Question: where would you look for a symbiotic soil fungus that functions well at high temperatures? The answer came with the discovery of a heat-tolerant fungus associated with the roots of plants growing near geothermal sites in the United States. Experiments with the fungus *Curvularia sp.* have shown that plants with roots inoculated with the fungus grow far better in exposed restoration sites than those with no root fungus. The prospect of creating thermotolerant plant species through the manipulation of symbiotic fungi in their roots is a major advance for the restoration industry.

Since we first wrote in Chapter 9 about biological monitoring, there have been significant advances in the use of invertebrates to detect and monitor environmental change. Several laboratories around the world are now using not just one or two species but hundreds. Sampling many different kinds of invertebrates provides insights into the status of many aspects of the local ecosystem. When comparing changes brought about by different logging practices, for example, a large amount of information on the decomposer, herbivore, and predator communities present can be derived if samples are taken from several sites, including a reference site, with each sample containing two or three hundred species of beetles, flies, ants, butterflies, and spiders. Broad animal and vegetation sampling makes it possible to track changes in major components of the biodiversity and, when these changes appear destructive, to suggest ways of modifying the human activity so as to restore the lost elements. The use of hundreds of species for biomonitoring relies on some fancy computing, but it is increasingly successful in monitoring the impacts of different forestry practices, detecting river pollution, and measuring the progress of restoration projects. An intriguing product of the sampling is the creation of virtual invertebrate collections—that is, files of high-quality digital images of the invertebrates being monitored. Thus a reference collection of a few thousand invertebrate specimens no longer occupies a musty museum vault of alcohol-filled vials or fills drawers of pinned insects; instead, it fits on a small section of a hard drive or a single compact disc.

More Blueprints: Robotics and Adhesives

We end our catalogue of recent wild solutions with a brief collection of some new and striking developments in robotics and adhesives. In Chapter 11 we discussed the contributions of a wide range of animal types to the exciting field of robotics. Since then studies of animal visual guidance systems have led to the manufacture of self-navigating robots and unpiloted airborne vehicles (UAVs). Professor Mandyam Srinivasan and his team have used the visual and navigation systems of fast-flying insects to understand how a robot or UAV might achieve the precision and speed of a fly landing on a twig or a bee swiftly but unerringly finding a series of flowers. These animal models have enabled researchers to build a self-navigating helicopter designed for surveillance missions too dangerous to risk the life of a human pilot; a self-guided, object-avoiding robot; and a system in which a UAV can approach a target while appearing to be stationary—an old dragonfly trick.

Our brief review in Chapter 10 treated adhesives and antifouling chemicals from nature separately, but among the many developments in these fields in the past two years, one exciting project has brought them together. We talked about the glue that mussels use to cling to rocks on the wet and salty sea shore. Now a group of medical researchers has taken proteins from this glue to solve an extremely difficult problem. Metal surgical implants and catheters are frequently fouled by unwanted cell growth, bacteria, or blood clots, much like mussels and other invertebrates fouling the hulls of ships. The surprise of this research is that the adhesive proteins taken from the mussel are used to deliberately foul implants. The strategy works because the same molecules possess surface properties that can anchor medically safe, polymer antifouling layers; these, in turn, prevent fouling growths. In another twist to this story, the system may be further developed for antifouling paints for ship's hulls and manufactured marine surfaces. Further still, such powerful adhesives may even be applied to create robust deicing surfaces for aircraft wings.

Finally, what do flies, spiders, and geckos have in common? Each can crawl on walls and ceilings with ease. They achieve this by having fine hairs at the ends of their legs that fit into the minute cracks in the paint that the eye cannot see. Scientists have found in these "hairy attachment systems" an intriguing wild solution. Although the hairs end in a variety of shapes, most commonly a flattened pad, it has become clear that the heavier the animal, the smaller the pads and the more densely packed the hairs at the end of the leg. These are both linear relationships suggesting that these diverse animals have evolved a design principle: splitting the contact between the leg and the landing site into finer and finer contacts increases adhesion. Researchers are investigating the application of the principle to the development of self-cleaning, reattachable, dry adhesives. They might come in the form of tape; it appears, for example, that a pair of human palms covered with "gecko tape" would support a human body on a vertical surface. Spiderman beware.

All Species Matter

What about species (or populations) that don't seem to hold promise of providing us with new products like adhesives, medicines, or materials? Are they expendable? Should we show no concern when a population of a "useless" butterfly is buried under a shopping mall or when rapid climate change exterminates a species of tiny "useless" orchid? We think that every species should be a source of concern because reasons for conservation are not only utilitarian but ethical and esthetic as well. Earth's biodiversity comprises our only known living companions in a large and desolate universe. There may be life forms elsewhere, and we may be about to discover them, but even if there are species on other planets, can we afford to dispense with ours? Without attempting to probe deep philosophical questions, we look around and agree with those people who quite simply believe that we humans have some duty of care for the species that share our planet.

Beyond the ethical argument that all species matter is an esthetic one. We agree with the great French structural anthropologist Claude Lévi-Strauss, who stated that any insect species is "an irreplaceable marvel, equal to the works of art we religiously preserve in museums." Anyone who has looked closely at a butterfly, who has seen under a microscope a tiny bee or fly that looks as if it were carved from a solid chunk of gold, or the glassy lattice skeleton of a single-celled marine alga, or who has simply dissected the amazing tracery of an insect's silvery breathing tubes could not but agree. Birdwatchers, hikers, and scuba divers draw infinite esthetic pleasure from biodiversity, as do gardeners and aquarium enthusiasts. The tourism industry relies on the presence of intact ecosystems, bursting with animal and plant life. *Every* species adds to the beauty, complexity, and wonder of life—and as any butterfly collector can testify, so does the infinite variation in color and pattern from population to population of the same species.

This book has focused on the practical reasons to consider every species (and every population) at least potentially important. No single species or population has ever been completely examined for its potential value to humanity—indeed, as we have seen, as new tools become available, new uses for old species leap out at us. It seems unlikely that a gecko would have been considered a potentially important source of insight into adhesives before the invention of the electron microscope. Equally, the role that most species play in ecosystems is incompletely known. We don't know how many species that currently seem redundant have the potential to be rescuers of humanity on the scale of *Cactoblastis* or any of the other obscure species used for biological control (Chapter 7). But the *Cactoblastis* tale and similar stories tell us that we should be cautious about considering any species to be redundant. Especially with climate change speeding up, conditions may change so that organisms that are today redundant from a human viewpoint may be crucial tomorrow. An obscure wild relative of rice, for example, may contain precisely the gene for drought tolerance that allows dryland rice production to persist in crucial areas as the globe warms.

The species in our life-support systems have been likened to the rivets in an airplane's wing. Many may seem redundant, and the function of some is obscure. But if we make a habit of popping the rivets from our airplane's wing, sooner or later we're likely to pay a very high price for that behavior. It seems to us that the same applies to "popping" species from the structures—the ecosystems—that support us. It makes no sense to squander our biological capital.

The importance of that capital, as well as the flow of "interest" it provides in the form of ecosystem services, is now widely recognized by environmental and social scientists and is gradually becoming important to economists. In fact, these groups are collaborating on a massive global project, the Millennium Ecosystem Assessment (MEA), to assess the ability of Earth's ecosystems to continue to support humanity. Sponsored by various agencies in the United Nations and a wide variety of international organizations, hundreds of ecologists, earth scientists, economists, social scientists, and political scientists from all over the world will provide this assessment in a series of reports to be released in 2005. These reports will be written for national leaders and governments, as well as decision makers at the regional and local scales. They will include projections of alternative scenarios that depend on the choices we make now. The MEA is solid evidence that people all over the world are capable of making the changes required to preserve our biological wealth and our ability to continue finding wild solutions.

Sydney, Australia, 23 January 2004

For more about these topics, see the "Afterword" section of the following Recommended Reading.

Recommended Reading

Agosta, W. 1995. *Bombardier Beetles and Fever Trees.* New York: Addison-Wesley.

Altieri, M. A. 1994. *Biodiversity and Pest Management in Agroecosystems.* New York: Haworth Press.

Anderson, D. 1994. Red tides. *Scientific American* 271 (2):52–58.

Beattie, A. J. 1992. Discovering new biological resources—chance or reason? *Bioscience* 42:290–292.

Beattie, A. J. 1994. Natural history at the cutting edge. *Ecological Economics* 13:93–97.

Benyus, J. M. 1997. *Biomimicry.* New York: Quill William Morrow.

Bright, C. 1998. *Life out of Bounds: Bioinvasion in a Borderless World.* New York: W. W. Norton.

Buchmann, S. L., and Nabhan, G. P. 1996. *The Forgotten Pollinators.* Washington, D.C.: Island Press.

Carson, R. 1962. *Silent Spring.* Boston: Houghton Mifflin.

Chichilnisky, G., and Heal, G. 1998. Economic returns from the biosphere. *Nature* 391:629–630.

Clutton-Brock, J. 1999. *A Natural History of Domesticated Mammals.* Cambridge, England: Cambridge University Press.

Cohen, J. E., and Tilman, D. 1996. Biosphere 2 and biodiversity: The lessons so far. *Science* 274:1150–51.

Colborn, T., Dumanoski, D., and Myeres, J. P. 1996. *Our Stolen Future.* New York: Abacus.

Connif, R. 1996. *Spineless Wonders.* New York: Henry Holt.

Crawford, R. L., and Crawford, D. L. *Bioremediation: Principles and Applications.* Cambridge, England: Cambridge University Press.

Daily, G. C., ed. 1997. *Nature's Services.* Washington, D.C.: Island Press.

Deacon, J. W. 1997. *Modern Mycology.* Oxford: Blackwell Science.

Edwards, C. E. 1998. *Earthworm Ecology.* New York: CRC Press.

Ehrlich, P. R., and Ehrlich, A. H. 1981. *Extinction: The Causes and Consequences of the Disappearance of Species.* New York: Random House.

Gaston, K. J., and Spicer, J. I. 1998. *Biodiversity: An Introduction.* Oxford: Blackwell Science.

Gordon, D. 1999. *Ants at Work.* New York: Free Press.

Hellin, J., Haigh, M., and Marks, F. 1999. Rainfall characteristics of Hurricane Mitch. *Nature* 399:316.

Heuer, A. H., Fink, D. J., Laraia, V. J., Arias, J. L., Calvert, P. D., Kendall, K., Messing, G. L., Blackwell, J., Rieke, P. C., Thompson, D. H., Wheeler, A. P., Veis, A., and Caplan, A. I. 1992. Innovative materials processing strategies: A biomimetic approach. *Science* 255:1098–1105.

Hillel, D. 1991. *Out of the Earth: Civilization and the Life of the Soil.* New York: Free Press.

Hokkanen, H. M. T., and Lynch, J. M., eds. 1995. *Biological Control: Benefits and Risks.* Cambridge, England: Cambridge University Press.

Hölldobler, B., and Wilson, E. O. 1990. *The Ants.* Cambridge, Massachusetts: Belknap Press of Harvard University Press.

Julien, M. H., and Griffiths, M. N. 1998. *Biological Control of Weeds.* Wallingford, Connecticut: CABI Publishing.

Knutson, R. M. 1999. *Fearsome Fauna: A Field Guide to the Creatures That Live in You.* New York: W. H. Freeman.

Lewis, W. H., and Elvin-Lewis, M. P. F. 1977. *Medical Botany.* New York: Wiley-Interscience.

Louda, S. M., Kendall, D., Conner, J., and Simberloff, D. 1997. Ecological effects of an insect introduced for the biological control of weeds. *Science* 277:1088–90.

Lowman, M. D. 1999. *Life in the Treetops.* New Haven: Yale University Press.

Margulis, L., and Schwartz, K. V. 1998. *Five Kingdoms: An Illustrated Guide to the Phyla of Life on Earth.* New York: W. H. Freeman.

Myers, N. 1983. *A Wealth of Wild Species.* Boulder, Colorado: Westview Press.

Naylor, R., and Ehrlich, P. R. 1997. Natural pest control services and agriculture. In G. C. Daily, ed., *Nature's Services,* pp. 151–174. Washington, D.C.: Island Press.

Pielou, E. C. 1998. *Fresh Water.* Chicago: University of Chicago Press.

Pimentel, D., and Levitan, L. 1988. Pesticides: Where do they go? *Journal of Pesticide Reform* 7:778–784.

Plotkin, M. J. 1993. *Tales of a Shaman's Apprentice.* New York: Penguin Books.

Postel, S. L., Daily, G. C., and Ehrlich, P. R. 1996. Human appropriation of renewable fresh water. *Science* 271:785–788.

Postgate, J. 2000. *Microbes and Man.* Cambridge, England: Cambridge University Press.

Powers, A. 2000. *Nature in Design.* Octopus.

Root-Bernstein, R., and Root-Bernstein, M. 1997. *Honey, Mud, Maggots, and Other Medical Marvels.* London: Macmillan.

Rosebury, T. 1969. *Life on Man.* London: Paladin Press.

Schneider, S. H. 1997. *Laboratory Earth: The Planetary Gamble We Can't Afford to Lose.* New York: Basic Books.

Schultes, R. E., and Raffuaf, R. F. 1990. *The Healing Forest.* Dioscorides Press.

Smith, B. L., Schaffer, T. E., Viani, M., Thompson, J. B., Frederick, N. A., Kindt, J., Stucky, G. D., Morse, D. E., and Hansma, P. K. 1999. Molecular mechanistic origin of the toughness of natural adhesives, fibres and composites. *Nature* 399:761–763.

Tsui, E. 2000. *Evolutionary Architecture.* New York: John Wiley.

Uphof, J. C. T. 1968. *Dictionary of Economic Plants.* Codicote, England: Wheldon and Wesley.

Vitousek, P. M., Aber, J. D., Howarth, R., Likens, G., Matson, P. A., Schlesinger, W., Schindler, D., and Tilman, D. 1997. Human alteration of the nitrogen cycle. Ecological Society of America, *Issues in Ecology,* no. 1.

Vitousek, P. M., Mooney, H. A., Lubchenco, J., and Melillo, J. M. 1997. Human domination of Earth's ecosystems. *Science* 277:494–499.

Walters, D., and Proctor, H. 1999. *Mites: Ecology, Evolution and Behavior.* New York: CABI Publishing.

Willis, D. 1995. *The Sand Dollar and the Slide Rule.* New York: Addison-Wesley.

Wilson, E. O. 1992. *The Diversity of Life.* Cambridge, Massachusetts: Belknap Press of Harvard University Press.

Zill, S. N., and Seyfarth, E. A. 1996. Ekoskeletal sensors for walking. *Scientific American* (July).

Afterword

Arzt, E., Gorb, S., and Spolenak, R. 2003. From micro to nano contacts in biological attachment devices. Proceedings of the National Academy of Sciences 100:10603–10606.

Coley, P. D., Heller, M. V., Aizprua, R., Arauz, B., Flores, N., Correa, M., Gupta,

M., Solis, P. N., Ortega-Barria, E., Romero, L. I., Gomez, B., Ramos, M., Cubilla-Rios, L., Capson, T. L., and Kursar, T. A. 2003. Using ecological criteria to design plant collection strategies for drug discovery. *Frontiers in Ecology and Environment* 1:421–428.

Crookes, W. J., Ding, L. L., Huang, Q. L., Kimbell, J. R., Horwitz, J., and McFall-Ngai, M. J. 2004. Reflectins: The unsusual proteins of squid reflective tissues. *Science* 303:235–238.

Daily, G. C., and Ellison, K. 2002. *The New Economy of Nature.* Washington, D.C.: Island.

Delcomyn, F. 2003. Insect walking and robotics. *Annual Review of Entomology* 49:51–70.

Demain, A. L. 2000. Small bugs, big business: The economic power of the microbe. *Biotechnology Advances* 18:499–514.

Ehrlich, P. R., and Ehrlich, A. 2004. *One with Nineveh: Politics, Consumption and the Human Future.* Washington D.C.: Island.

Frankham, R., Ballou, J. D., and Briscoe, D. A. 2002. *Introduction to Conservation Genetics.* Cambridge, England: Cambridge University Press.

Laird, S. A. 2002. *Biodiversity and Traditional Knowledge.* London: Earthscan.

Mann, S., Webb, J., and Williams, R. J. P. 1989. *Biomineralization.* New York: VCH.

Millennium Ecosystem Assessment. 2003. *Ecosystems and Human Well-Being.* New York: Island.

Sarikaya, M., Tamerler, C., Jen, A. K., Schulten, K., and Baneyx, F. 2003. Molecular biomimetics: Nanotechnology through biology. *Nature Materials* 2:577–585.

Wakeford, T. 2001. *Liaisons of Life: How the Unassuming Microbe Has Driven Evolution.* New York: John Wiley.

WEBSITES

Environmental monitoring: www.biotrackaustralia.com.au

Lotus Effect: www.lotus-effect.com

Millennium Ecosystem Assessment: www.millenniumassessment.org

Robotics: http://biology.anu.edu.au/bioroboticvision/

Biomimetics: http://www-cdr.stanford.edu/biomimetics/

Index